From Abundance to Function: Exploring the Complex World of Membrane Proteins

Nama

Table of contents

1 Introduction

Membrane proteins play an important role in cellular function as receptors, transporters, anchors and catalysts. About 20-30% of the human proteome is associated with cell or organelle membranes by at least one membrane spanning domain,[1] yet, only 3 % of characterized proteins are membrane proteins.[2] Tendency to aggregate when isolated outside of a lipid bilayer is only one of the many challenges researchers are faced with when working on membrane proteins. To access the multitude of biophysical and biochemical factors that drive protein-lipid and protein-protein interactions inside of biological membranes, it is helpful to break down research targets into well-defined and well characterized model systems.[3,4] This includes the need for transmembrane peptides that can mimic certain properties as polarity, dipole, geometric characteristics or featuring whole sequence fragments, while still preserving the ability to insert into a lipid bilayer.[5–7] Peptides and proteins with natural amino acid composition can be available by recombinant techniques with low error rate, courtesy of the well-functioning DNA-translation and transcription machinery in live cells.[8] When artificial modifications are required however, enzyme scope is usually exceeded while chemical peptide synthesis – although being vulnerable to errors in synthesis – excels. Solid phase peptide synthesis (SPPS) can accommodate a wide array of modifications if they are compatible with synthetic conditions of SPPS.[9,10] With no correction mechanisms in place however, the achievable length is limited by the yield of each reaction step, as errors accumulate exponentially over the course of the synthesis. While for soluble peptides the desired product can often be easily isolated even from an excess of byproducts by high performance liquid chromatography (HPLC), transmembrane peptides tend to be less forgiving, complicating commonly used purification techniques with solubility issues and emphasizing the importance of good quality crude product.[11,12]

SNARE-mediated (soluble *N*-ethylmaleimide-sensitive factor attachment protein receptor) fusion of biological membranes can be found in a wide array of organisms ranging from yeast over plants to mammals, including humans.[13] The highly regulated process essential to transport of material across biomembranes has long been under thorough investigation and since the first characterization of SNARE proteins in the late 80s their central role in the secretory pathway of eukaryotic cells is undisputed.[14,15] For neuronal exocytosis complementary SNARE proteins syntaxin-1A and synaptobrevin-2 are located in the membranes of transmitter-filled synaptic vesicles and the presynaptic membrane, respectively, anchored by transmembrane domains.[16] Together with 25 kDa synaptosome-associated protein (SNAP 25), which is attached to the plasma membrane by a lipid anchor, upon contact, the soluble domains form a tight four-helix bundle, pulling the opposite membranes into close proximity. SNARE complex formation is followed by membrane fusion and transmitter release.[16] A plethora of model fusogens have been developed to imitate the structure and function of the membrane associated proteins, most focusing on the recognition units tethering the opposite membranes together.[17,18] Instead of transmembrane domains, many fusion protein mimetics use lipid anchors or other hydrophobic organic molecules to secure the fusogens inside the lipid bilayers. Recognition principles range from selective small molecule interactions over

DNA duplex formation to structurally SNARE-motif-related coiled coil forming peptides E3 and K3.[19–22] These model systems are usually limited to the aspect of membrane approximation achieved by the recognition units but only superficially address the role of the transmembrane domain reducing it to its anchoring function. *De novo* design of fusogenic transmembrane peptides has shown that conformational flexibility of the membrane spanning peptide achieved by mixed Leu/Val sequences is beneficial for fusion efficiency.[23,24] Furthermore, amino acid exchange in synaptobrevin-2 was found to modulate fusion rates, accelerating fusion pore formation when isoleucine and valine content was increased and slowing it down with α-helix-stabilizing leucine.[25] This highlights that exceeding the hydrophobic properties of the residues, amino acid sequence of the transmembrane section impacts the fusion process on a molecular level.

In the DIEDERICHSEN group, a model system has been established, that combines coiled coil forming peptides E3 and K3 with the transmembrane domain and linker region of neuronal SNAREs synaptobrevin-2 (Syb) and syntaxin-1A (Sx) to obtain E3Syb and K3Sx.[26] The topologically homogeneous fusogens allow to address the zippering hypothesis, a potential key aspect of membrane fusion in which dimerization starts at the *N*-terminus of the SNARE motif and continues beyond the linker into the transmembrane region in a zipper like fashion.[27,28] The peptidic model fusogens were shown to induce full fusion in large lamellar vesicles by bulk FRET based fusion assays.[26] One objective of this work is to reversibly halt vesicles reconstituted with E3Syb and K3Sx in a docking or hemifusion state and restart the process with a fast and clean trigger. Photolabile protecting groups (PPGs) are a popular tool to provide spatial and temporal control in biochemical processes and therefore ideal modulators for this project.[29] Targeting the membrane-proximal region of the coiled coils, a photoprotection strategy is developed to disrupt the attractive interactions between the heterodimer forming pair by spanning sterically demanding PCGs through the hydrophobic core of the coiled coil. Inspired by all-hydrocarbon stapled peptides developed in the VERDINE group,[30,31] photocleavable units equipped with allyl linkers are incorporated into peptide E3 at different distances and bridged by Grubbs I catalyzed ring closing metathesis (RCM). The new PPG is characterized by spectroscopic, chromatographic and mass spectrometry methods. A selection of unprotected and photocleavably protected E3Syb and K3Sx derivatives are synthesized and compared regarding their fusogenicity. Fusion activity is assessed by bulk FRET based fusion assays in large unilamellar vesicles (LUVs).

Inter- and intracellular trafficking is likewise central for the activation of adaptive immune responses. For a cytotoxic T cell to be able to destroy potentially dangerous cells it has to be activated by recognizing short peptide fragments of about 9 amino acids on the surface of dendritic cells (DCs).[32,33] The task of DCs is to scan their surroundings for signs of disease by taking up pathogens and cell material. After processing their findings, DCs present short peptide fragments of the digested proteins on their surface on major histocompatibility complex I (MHC-I) in a process called cross-presentation. Recognition of the epitopes by naïve CD8$^+$ T cells prompts cellular immune response. For antigens originating from soluble proteins the mechanisms leading to an immune response are well understood.[34] Two principal pathways, the phagosome-to-cytosol (P2C) pathway and the vacuolar pathway, that are considered to be majorly responsible for cross-presentation have been carved out in numerous studies.[35–38] However, very little research is directed towards the immune response activated by membrane buried epitopes[39] and the known pathways have not yet been shown to include crucial steps that would be required to extract, process and transport epitopes from membrane proteins.

Artificial peptides have been designed in the VAN DEN BOGAART[1] group to contain known tumor epitope NY-ESO1[157-165] flanked by hydrophobic amino acids to form an alpha helix inside lipid bilayers. The epitope is equipped with a small bio-orthogonal linker to bear the potential for fluorescent labeling for intracellular tracking without interference with peptide processing. A variety of peptides with different positioning of the epitopes within the transmembrane helix and different amino acids of the epitope substituted for propargylglycine ({pra}) are synthesized, reconstituted into liposomes and scanned for being processed by monocyte derived dendritic cells isolated from human donor blood. Cross-presentation is verified by quantification of cytokine interferon γ (INFγ) secreted by CD8$^+$ T cells upon activation. Employing pharmaceuticals that selectively inhibit metabolic processes involved in cross-presentation, suitable peptides are then utilized to gradually dissect cross-presentation of membrane-buried epitopes. Additionally, the {pra} linker is exploited in soluble antigens as an alternative assay for quantifying cross-presentation by fluorescent labeling developed for mouse models by PAWLAK *et al.*[40] The assay is translated to use on primary human cells with clinically relevant epitopes and adapted in the study of phosphoinositide kinase PIKfyve involvement in cathepsin-S mediated MHC-II antigen presentation in human derived dendritic cells.[41] Peptide synthesis and purification are performed as part of the scope of this book. All peptide designs and *ex vivo* studies are executed by FRANS BIANCHI, ELKE MUNTJEWERFF, MAKSIM BARANOV and SJORS MAASSEN.[1]

The research targets taken on in this work are partially connected by synthetic and chromatographic challenges related to difficult peptides which are solved similarly across projects. Otherwise, the motivations, synthetic approaches and used techniques are specific to the distinct project. Therefore, the following sections will generally introduce to membrane fusion and pathways of antigen presenting cells. Motivations and objectives will then be presented in the beginning of each chapter.

---

[1] Radboud Institute for Molecular Life Sciences in Nijmegen and Groningen Biomolecular Sciences and Biotechnology Institute

1.1 Membrane fusion

Biomembranes serve as physical boundaries to cells but they also constitute essential elements of their interior providing spatially and functionally defined reaction chambers known as organelles. Lipids, which assemble into a bilayer of approximately 5 nm thickness form the main structural feature of membranes. Additionally, they are crowded with proteins, which occupy about 50% of the plasma membrane surface[42] and biomolecules such as saccharides are displayed on cell surfaces as essential markers for cell-cell recognition.[43] Far from only being passive separating layers they are important for many other functions such as generation of energy, signaling, and directed transport of material.

Lipid bilayers, composed of large numbers of individual and diverse lipid molecules, owe their high structural integrity to the amphipathic nature of their components. The polar headgroups facing outwards shield the hydrophobic tails on the inside from entropically unfavorable interactions with water molecules. At the same time, the bilayers are subject to three main strains – stretching, tilt, and curvature[44] – which are in turn influenced by osmotic pressure between the enclosed and surrounding medium, interactions with membrane proteins and adjacent proteins, and adhesion forces to surfaces.[45] How well the bilayers can respond to these strains, i.e. how much they can take and still maintain their integrity, is majorly determined by the preferred arrangement of the given lipid composition (see Figure 1.1).[46,47] For instance, lipids with the headgroup phosphatidylcholine (PC) such as dioleylphosphatidylcholine (DOPC) typically have a cylindric molecular shape and favor planar architectures. Cone-shaped lipids (e.g. dioleylphosphatidyletanolamine (DOPE)) and inverted-cone-shaped lipids have a negative and positive intrinsic curvature, respectively.[46] Membranes containing cholesterol have an increased cohesion which is expressed in increased stiffness and decreased permeability to water.[48–50]

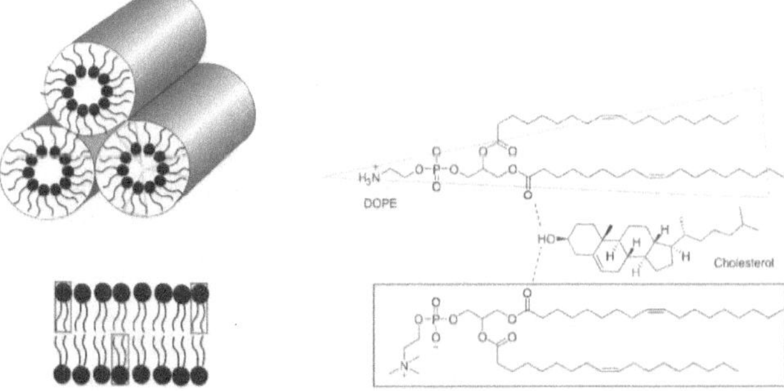

While maintaining the macroscopic structure, lipids (and to some extent other membrane components) are still allowed to diffuse and deform laterally (fast) and across the bilayers (known as flip-flop[51], slow). In the Singer-Nicolson fluid mosaic model first published in 1972, lipids, proteins, and other membrane components were described to be homogeneously distributed in the membrane.[52] However, observations of phenomena like hierarchically built supramolecular protein complexes,[53,54] restricted diffusion of membrane-proteins[55,56], and domains with distinct lipid compositions termed "rafts"[57] have shaped a new understanding of the structure of biomembranes. A modern interpretation views membranes to be highly structured and transversely asymmetrical, with the cytosolic side of plasma membranes supported by a skeleton of cortical actin.[58] How the high degree of organization is achieved and maintained is not entirely understood, but an interplay of physical factors such as membrane curvature and lipid miscibility, cytoskeletal interaction, and enzymatic interventions e.g. through ATP-requiring flippases appears plausible.[58,51]

Spatially and temporally controlled exchange of material is a fundamental task of biomembranes that must be reconciled with the effective separation of cells and their compartments. As one of several transport mechanisms, membrane trafficking transfers biomolecules through small lipid vesicles budded from the membrane of the source compartment. Large and polar cargo is carried inside the volume enclosed by the vesicles and released by mixing of the volumes with the destination compartment. Besides, membrane-bound components like ion channels, enzymes, and lipids can be transported to a target membrane and become part of it. Membrane fusion between the vesicle and target membranes is a highly regulated process. Diverse proteins facilitate the correct recognition of the membranes destined to fuse and approximate them to where their intrinsic stability is overcome, and lipids start to mix. Proteins of the SNARE-family have been recognized as fusion mediators in physiological fusion processes involved in intracellular transport and neurotransmitter release in neuronal exocytosis.[59–61] Pathological fusion events as found in viral infection typically require less regulation and precision and are often promoted by a single type of protein e.g. hemagglutinin (HA) of influenza A.[60]

1.1.1 Mechanics of membrane fusion

When two biomembranes merge this leads to mixing of their lipids as well as of the enclosed volumes, in other words, two previously separate lipid bilayers become one continuous bilayer with a single enclosed volume. Before this can happen, a steep energy barrier must be overcome. Dipolar headgroups of the bilayers interact strongly with water molecules which must be displaced in order to bring the bilayers into close enough distance to fuse. The actual energetic cost of membrane fusion is highly dependent on the lateral membrane tension (strains acting on the membrane) at the fusion site.[62,45,63] In living organisms, the energy needed for fusion is usually raised by specialized fusion proteins, however, the pathways by which a connection between two membranes is formed are more extensively studied in protein-free models both theoretically and experimentally.

Depending on the used methods and settings, the suggested sequences of events differ from each other (see Figure 1.2) and it is conceivable that fusion pathways may be dependent on the applied conditions e.g. the lipid composition. Two stages of fusion are commonly agreed upon: The first connection between the two membranes is referred to as the fusion stalk (C). This metastable intermediate is formed from lipids of the outer (*cis*) layers of the two membranes destined to fuse. Although usually short-lived, the stalk can become a stable structure under the right conditions like low hydration and negative spontaneous curvature and is commonly identified in X-ray studies though the rhombohedral diffraction patterns of the hexagonal phase in stalk lipids.[64,65] A late stage that establishes an aqueous connection between the two compartments is accepted to involve at least one fusion pore, formed by lipid rearrangement of the inner (*trans*) layers into a shared continuous layer (E and G).

KOZLOV and MARKIN first proposed the so-called stalk model in 1983 which would shape the understanding of membrane fusion in the following decades.[66] With the initial assumptions concerning the shape of the fusion stalk the energy needed for fusion was vastly overestimated so the model was overhauled to allow the tilting of lipids for a better approximation of the energy cost in biological membrane fusion with about 40 $k_B T$.[67–69] The stalk model of membrane fusion – or fusion-through-hemifusion as denoted by the authors[70] – follows a continuum approach model and is described as a series of the following steps: When the apposed membranes come into close enough distance (A), a thermally powered point-like protrusion (B) transforms into the axially symmetric hemifusion stalk (C). From here, the stalk can radially expand into a hemifusion diaphragm (D) – formed from the originally *trans* monolayers – which then ruptures, or directly forms the fusion pore (E). Both options assume accurate separation between the lipids of the *cis* and *trans* layers and also fusion without content leakage. Hemifusion diaphragms[71] have been verified experimentally, however the direct progression from diaphragm to fusion pore has been questioned because of the high stability of the intermediate.[72]

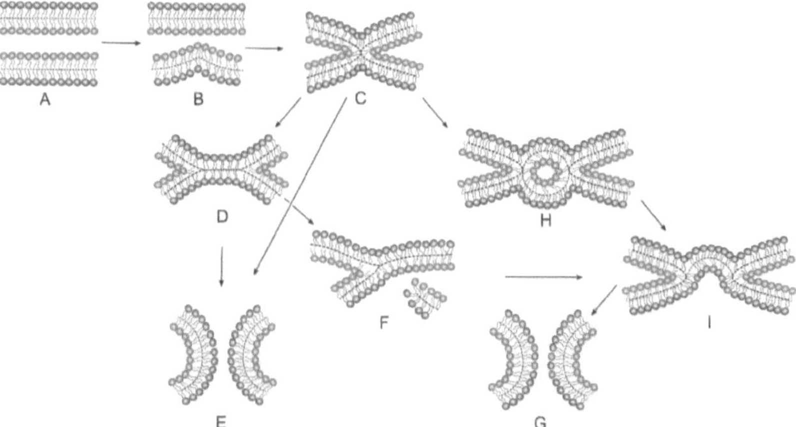

Figure 1.2: Possible pathways of protein-free membrane fusion, annotations in the text. Adapted in accordance with CHERNOMORDIK *et al.*[70] *and* FUHRMANS *et al.*[73]

Corse grained simulations have offered explanations on how less neat fusion could take place. Passing the stalk (C) and hemifusion diaphragm (D), fusion-through-rupture (F→I→G) has been proposed.[74,75] This route allows the leakage of content and mixing of *cis* and *trans* lipids which is sometimes observed in viral membrane fusion. Another pathway to a *cis* and *trans* mixed fusion pore has been published by RISSELADA *et al.*, demonstrating a longitudinal elongation of the fusion stalk (C), which then collapses on itself to encapsulate a small volume in an inverted micelle intermediate (H).[76] Two pores must be opened to connect all compartments, one to create a hemifusion diaphragm-like intermediate (I) and the final fusion pore (G).

In studies on protein mediated membrane fusion, the concepts described above are often embraced. This was supported by the finding, that a single set of SNARE-proteins suffices to induce membrane fusion *in vitro*.[77] Within the assumed transitions between intermediates, the proteins are thought to bring the membranes into close distance and to support the rearrangement of lipids by perturbing lipid packing. Modulation of membrane curvature by peptides is also considered a contributing factor.[78] In the controversial transition from hemifusion diaphragm to fusion pore – with diaphragms having been visualized in protein-mediated fusion[79] – the proteins may restrict the lateral expansion of the diaphragm, thus limiting its stability and promoting the opening of a pore.[80] However, a different concept has emerged in the early 90s which imagines a direct involvement of the fusion proteins in the formation of the fusion pore.[81,82] A proteinaceous fusion pore, similar to gap-junctions and ion channels, was proposed to form the initial aqueous connection between two compartments and precede the mixing of lipids. Only gradually the pore-forming proteins drift apart, allowing lipids to line the fusion pore and expand it.[83,84] This concept was supported by the observation that aqueous connection preceded mixing of lipids in hemagglutinin (viral) promoted cell membrane fusion by several minutes.[85,86] It is not entirely clear how the hydrophobic transmembrane domains of fusion proteins could form a pore that allows polar water molecules to traverse. Lately, the interaction of transmembrane and juxtaposing protein domains with the membrane lipids has come into focus which was conceptualized in the image of a proteolipidic fusion pore.[87–89] So far, no concept was proven universally superior to the others. It is possible that several mechanisms are biologically relevant considering that fusion proteins act under diverse requirements for precision (viral vs. exocytosis) and temporal control (cell-cell fusion vs. neuronal exocytosis).

1.1.2 Neuronal exocytosis

Chemical neurotransduction in nerve terminals of vertebrate neurons is one of the most studied physiological events in the context of membrane fusion and fusion proteins. In their pioneering works, KATZ and coworkers have laid the foundation for uncovering the pathways of synaptic transmission as early as the 1950s and 60s.[90–92]

The connection between two nerve endings is called synapse which is where neuronal exocytosis occurs (Figure 1.3). Signals arriving at the nerve terminal as action potentials are passed on to the postsynaptic neuron by release of neurotransmitters into the synaptic cleft. From there, they contribute to building an action potential in the next neuron. To transfer the neurotransmitters across the presynaptic membrane, they are enriched inside of synaptic vesicles during resting phases. When the action potential arrives, voltage-gated Ca^{2+}-channels enable a transient Ca^{2+}-influx which in turn triggers fusion of the synaptic vesicles with the presynaptic membrane within one millisecond.[16] In the synaptic cleft the neurotransmitters initiate building of the next action potential. Fusion pore dynamics influence the number of released neurotransmitters. Initially only few nm wide, the pore can flicker open and closed multiple times before resealing (kiss and run) or committing to full dilation.[93] Recently, this behavior has been identified as an intrinsic aspect of membrane fusion.[94] Full merger of the vesicles causes the increase of surface area of the presynaptic membrane, but this change is not permanent. Membrane lipids and proteins are recycled into new vesicles by clathrin-dependent endocytosis.[95]

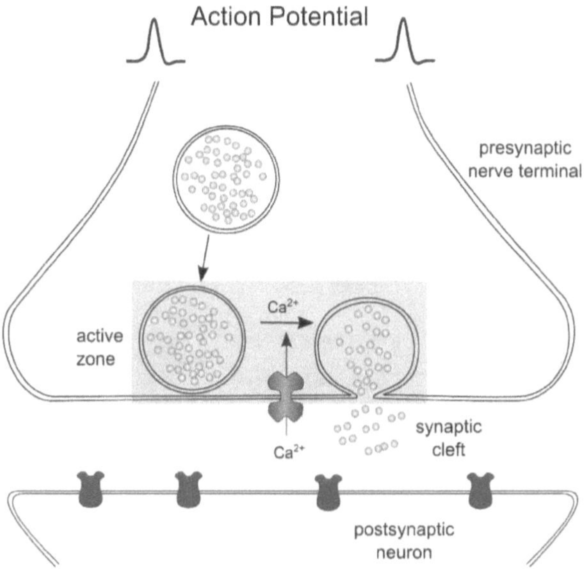

Figure 1.3: Schematic illustration of the synapse demonstrating the principles of chemical neurotransmission. Vesicles filled with neurotransmitter molecules are mostly found in the cytoplasm, but some are translocated to the active zone of the presynaptic membrane to be primed for fusion. The incoming action potential opens voltage-gated calcium channels and the influx of Ca^{2+} triggers synchronous fusion of all primed vesicles within < 1 ms. In the synaptic cleft neurotransmitters travel to the postsynaptic membrane and interact with receptors to build a new action potential. Based on JAHN et al. [95]

While being part of the secretory pathway and sharing multiple regulatory protein families with other intracellular processes involving membrane fusion, neuronal exocytosis has always stood out for its impressive performance regarding spatio-temporal control and speed. Particularly, neuronal SNARE proteins syntaxin-1A, SNAP-25 and synaptobrevin-2 – being identified as key fusion mediators and minimal fusion machinery[96] – have been in the center of countless studies e.g. structural and biophysical characterizations and cell free assays. Furthermore, they have inspired the structures of several synthetic model fusogens.

1.1.3 Neuronal SNARE proteins

SNARE proteins catalyze membrane fusion along the secretory pathway in eukaryotic cells. More than 120 distinct representatives of this highly conserved protein family have been discovered in animals, plants and fungi.[13] Inarguably, the most characteristic feature of all SNAREs is a stretch of 60-70 amino acids organized in heptad repeats known as the SNARE motif. Four SNARE motifs (from different SNARE protein groups) recognize each other to form a parallelly aligned tetrameric coiled coil – the SNARE complex – and induce membrane fusion. The attraction between the helices is mostly based on hydrophobic interactions (15 layers in total), except for the central – "zero" – interaction layer which is formed by polar interactions, almost always between three glutamine (Q_a, Q_b, Q_c) and one arginine (R) residues. This has also inspired the classification as Q- and R-SNAREs (see Figure 1.4 A and B).

Neuronal SNAREs Synaptobrevin-2, syntaxin-1A and SNAP-25 (25 kDa synaptosome associated protein) have become known as the minimal fusion machinery as a result of reconstitution experiments *in vitro*.[96] As few as one set of these proteins was demonstrated to be sufficient for observing membrane fusion.[77] Their simplified structure is illustrated in Figure 1.4 A.

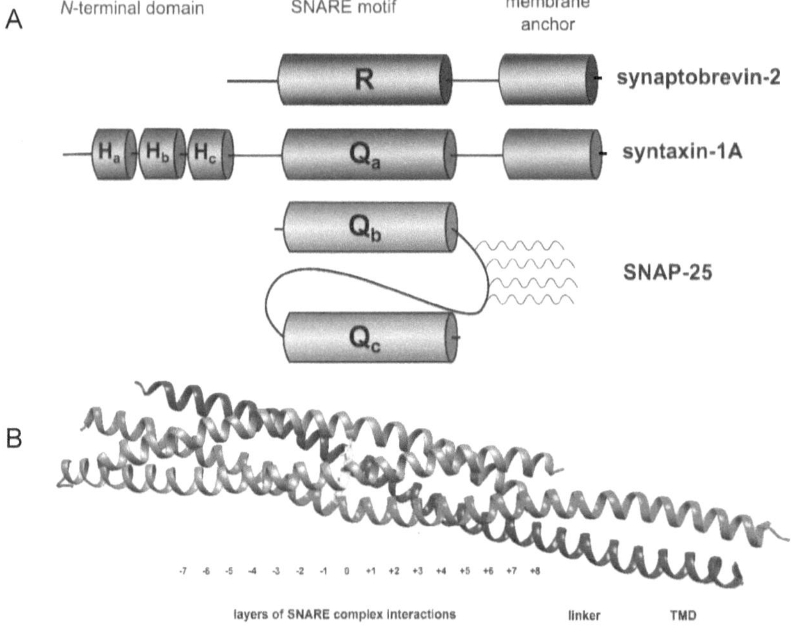

Figure 1.4: A) Schematic illustration of protein domains in neuronal SNARE proteins. Colored rods represent (mostly) helical structures while plain lines indicate unstructured sections and flexible linkers. Wavy lines depict palmitoyl anchors. Based on JAHN et al.[97]. B) Ribbon diagram of the SNARE complex formed by the SNARE motifs of synaptobrevin-2 (blue), syntaxin-1A (red) and SNAP-25 (green) and extended by the linker regions and TMDs of syntaxin-1A (red) and synaptobrevin-2 (blue) (PDB ID: 3H37).[98] Amino acids participating in the "zero" layer of

Synaptobrevin-2 is anchored in the synaptic vesicle by a transmembrane domain (TMD), earning it the alternative name VAMP (vesicle associated membrane protein) and the categorization as v-SNARE (vesicle). Syntaxin-1A (secured by a TMD) and SNAP25 (anchored by palmitoylation of cysteine residues) are known to be colocalized at the presynaptic membrane and are therefore classified as t-SNAREs (target). Syntaxin-1A additionally bears an N-terminal domain – a three-helix bundle known as H_{abc} domain extended by a shirt unstructured sequence referred to as N-peptide – which plays a role in the regulation of SNARE complex assembly (see section 1.1.4).[95]

When the SNARE complex is formed, synaprobrevin-2 and syntaxin-1A each contribute one α-helix whereas one SNAP-25 supplies two SNARE motifs. Assembly is thought to proceed in distinct zippering phases, beginning at the *N*-terminal end of the SNARE complex.[27] A partially zippered SNARE complex is assumed as a functional intermediate in docked vesicles (see section 1.1.4). Several biophysical studies have independently characterized the half zippered intermediates, all suggesting that the first stage of assembly halts at the -1 or 0 layer.[99,28,100] Progression to the second stage of SNARE complex zippering was measured to release a significant amount of energy – 36 k_BT – coming in the range of energy cost for biological membrane fusion.[99,69] However, the fully zippered complex only approximates the v- and t-membranes up to 2-3 nm, too far for spontaneous fusion stalk formation (~1 nm).[99,65] Furthermore, arrest of zippering just short of full SNARE complex formation by mutation in the +8 layer was shown to halt giant unilamellar vesicles in a tightly docked or hemifused state without progression to fusion.[79] LINDAU and coworkers suggested that the contribution of SNAREs to membrane fusion may exceed the interactions of the SNARE motifs and that zippering could continue via the linker regions down to the TMDs of synaptobrevin-2 and syntaxin-1A.[87,88] The progression of zippering would bring the membranes closer together and the assumed movement of charged *C*-termini inside the membrane was proposed to destabilize lipid interactions, lowering the energy barrier for stalk formation.[87] This vision is supported by a crystal structure obtained of the neuronal SNARE complex with linker regions and TMDs attached (Figure 1.4). Notably, the TMD-bearing SNAREs exhibit a continuous α-helical structure whereas their monomers were shown to be unstructured in the linker region at conditions where the SNARE motifs at least partially exhibit α-helical structure.[101,102] The depicted arrangement is suggested to be representative of the post-fusion state of the so-called *cis*-SNARE complex, indicating the fusion stage when v- and t-SNAREs are localized in the same membrane.[95]

The linker regions connecting SNARE motifs and TMDs are thought to play an important role as hinges, ensuring that the work performed by SNARE complex assembly is transmitted to the membrane. Increasing the length of linker regions was shown to inhibit fusion *in vivo* and *in vitro*.[103,104] Furthermore, the polybasic KARRKK sequence of syntaxin-1A seems to have multiple responsibilities. Interactions with the anionic lipid headgroups of the membrane might disrupt the organization of water molecules at the water/lipid interface.[105] This was proposed to facilitate dehydration between the apposing membranes and promote contact. Moreover, ionic bonds to specifically phosphatidylinositol-4,5-bisphosphate (PIP$_2$) play a role in the regulation of vesicle priming at the active zone of the presynaptic membrane (see section 1.1.4).

Presence of transmembrane domains is not mandatory to observe fusion, as illustrated by the numerous model fusogens with lipid anchors (see section 1.1.5).[17,18] This was even more impressively evidenced in a recent study by ZHOU *et al.* that demonstrated efficient Ca^{2+} triggered fusion in cultivated neurons using lipid anchored syntaxin-1A and synaptobrevin-2 mutants that lacked their TMDs. On the other hand, studies with altered SNARE TMDs commonly report a reduction in fusion efficiency compared to the native sequences. For example, in an inquiry by NGATCHAU *et al.* synaptobrevin-2 extension by even one lysine reduced fusion efficiency by 80%.[87] This surprising result was interpreted as evidence for fusion promoting interactions of the *C*-terminus with the inside of the bilayers which would be inhibited through anchoring of lysine to the polar headgroups. Lately, flexibility of secondary structure in TMDs has come into focus. After HOFMAN *et al.* presented *de novo* designed fusogenic peptides consisting of only a TMD, the design rules have been tested on synaptobrevin-2. Systematic exchange of amino acids in favor of either α-helix stabilizing leucin or β-sheet promoting isoleucine or valine were tested with regard to exocytosis efficiency. The extent of exocytosis could be positively correlated to the fraction of β-branched amino acids.[25] The notion, that TMDs might exist in a β-sheet conformation up until full SNARE complex assembly occurs, has been summarized as the β-to-α transition model, short BAT.[106] Transition to α-helical structure as a consequence of SNARE zippering is thought to promote lipid perturbations thus promoting the formation of a fusion pore. *In vitro* reconstitution of synaptobrevin-2 TMD in lipid multilayers could indeed verify the presence of β-sheet content between 7 and 53% depending in peptide-to-lipid ratio and lipid composition by ATR-spectroscopy.[107]

In conclusion, although the minimal fusion machinery has been identified over 20 years ago, the molecular mechanism by which it mediates membrane fusion is controversially discussed. Some of their structural features may only be understood in relation to regulatory mechanisms of Ca^{2+} dependent transmitter secretion.

1.1.4   Regulation of Ca^{2+} dependent neuronal exocytosis

Rapid response to an incoming action potential requires the preparation of some of the vesicles into a "readily releasable" state by a tightly cooperating ensemble of structurally conserved proteins. Docking and priming produces a pool of vesicles with partially assembled SNARE complexes at the active zone of the presynaptic membrane which can immediately respond to Ca^{2+} influx.[95]

Along with the SNAREs, the proteins regulating neuronal exocytosis belong to structurally conserved protein families. Docking and priming is guided mainly by four key regulators with potentially multiple functions: Sec/Munc18 proteins (SM, e.g. Munc18), CATCHR (in particular Munc13), complexins and synaptotagmins as Ca^{2+} sensors. A wealth of information on the roles of these proteins has become available through the research of the past thirty years shaping the understanding of the mechanism of neurotransmitter release.[108] Figure 1.5 summarizes one of the currently discussed sequences of protein interactions which leads to the correct assembly of the *trans*-SNARE complex.

Binding of Munc18 which contains three domains forming an arched-shaped structure to a self-inhibiting locked conformation of syntaxin-1A marks the beginning of the priming stage.[109,108] Thereby, the H_{abc} domain is folded over on the SNARE motif, preventing its interaction with other SNAREs. For a long time, the apparent inhibitory function of the SM protein seemed to contradict the complete loss of exocytosis in deletion mutants.[110] However, recent findings of HE *et al.* suggest, that Munc18 prevents the premature disassembly of SNARE complexes by NSF and α-SNAP.[111] Munc13, a large multidomain protein (200 kDa), has been suggested to contribute to opening the locked conformation. The notion that acceptor complex formation – consisting of syntaxin-1A and SNAP-25 – precedes docking of the synaptic vesicle has been widely accepted.[97],74] However, evidence is accumulating suggesting that the 1:1 complexes may not be a part of the assembly stage. Instead, Munc13 was suggested to promote vesicle docking by spanning a bridge between vesicles and target membrane and then guide parallel assembly of synaptobrevin-2 and syntaxin-1A SNARE motifs together with Munc18.[112,113,108] Only after, the two SNARE motifs of SNAP-25 are now thought to contribute to the SNARE complex. It is still unclear if Munc18-1 and Munc13 remain bound to the SNARE complex.

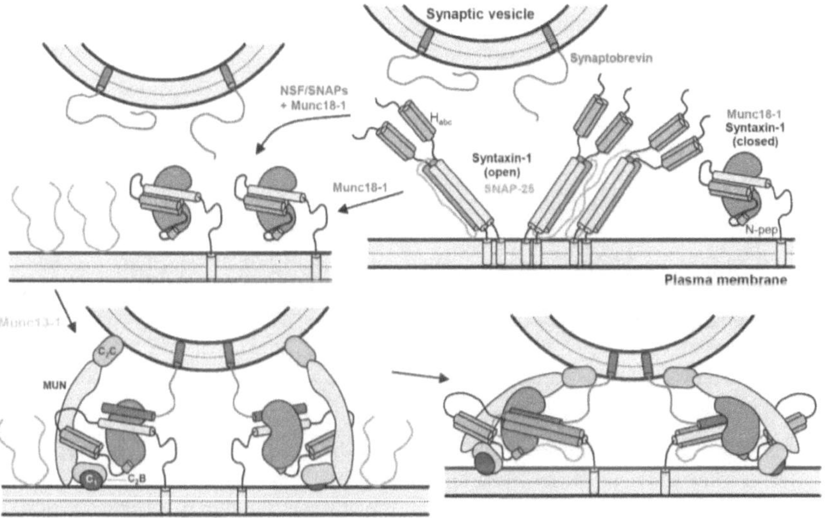

Figure 1.5: Recent concept of regulation of SNARE complex assembly. Starting from the top right and following the arrows: Munc18 binding to a closed conformation of syntaxin-1A initiates the assembly and prevents premature NSF/α-SNAP disassembly. Munc13 bridges the vesicular and the plasma membrane and releases the H_{abc} domain of syntaxin-1A. Munc18 guides the parallel arrangement of synaptobrevin-2 and syntaxin-1A SNARE motifs. In the last step, SNAP-25 joins to complete the trans-SNARE complex.[108]

Temporal control over membrane fusion likely is achieved through an interplay of complexins and synaptotagmins. The small protein complexin binds to the assembling SNARE complex in an antiparallel fashion, successively promoting *N*- to *C*-terminal zippering up to a certain point and then acting as a clamp – possibly through the *N*-terminal accessory helix – and arrests the *trans*-SNARE complex in a half-zippered state.[114] Thereby it reduces spontaneous transmitter release significantly yet doesn't completely eliminate it.[115] The calcium sensor synaptotagmin is eventually responsible for the synchronous response to Ca^{2+} influx. It is composed of a transmembrane domain anchoring it in the vesicle membrane and two C2 domains which can bind two and three Ca^{2+} ions, respectively. Several modes of action were proposed by which Ca^{2+}-dependent conformational and electrostatic changes in synaptotagmin translate to triggering membrane fusion. The complexin clamp is released upon Ca^{2+} influx which is thought to occur through interactions with the activated synaptotagmin, so that the SNARE complex can proceed to *C*-terminal zippering and unleash its full fusogenic potential.[116] The fusion efficiency may be enhanced by the activated synaptotagmin as it interacts with anionic phosphoserine lipids and even partially inserts into the plasma membrane, thus connecting vesicle and target membrane.[117] This way, it may aid in pulling the membranes closer together. Additionally, membrane penetration was suggested to induce a positive curvature, thereby reducing the energy barrier for fusion stalk formation.[118] Furthermore, an interaction between a polybasic lysine stretch of synaptotagmin and syntaxin-1A was proposed which is thought to occur via PIP_2 clusters arranged around the polybasic linker region of syntaxin-1A.[119]

When zippering of the SNARE complex is completed and membrane fusion has occurred, all SNARE proteins are located in the same membrane and now referred to as the *cis*-SNARE complex. The crystal structure depicted in Figure 1.4 B is widely accepted to represent the arrangement of the SNAREs at this point. The *cis* complex is highly stable and requires ATP consuming disassembly to be recycled. This task is accomplished by NSF (*N*-ethylmaleimide-sensitive factor) and up to four of its α-SNAP (soluble NSF attachment protein) cofactors. In a cryo-EM based study, recently the structure of the so-called 20S complex was reported.[120] It shows the SNARE four-helix bundle surrounded by four α-SNAP molecules, which are in contact with NSF at the *N*-terminal end of the SNAREs. The α-SNAPs were shown to twist around the SNAREs in the opposite direction to the left-handed supercoil. ATP hydrolysis is thought to power structural changes in NSF which is then translated to the α-SNAPs to unwind the SNARE complex. After disassembly, synaptobrevin-2 is endocytosed to be recycled in new synaptic vesicles.

SNAREs are the widely accepted fusion engines of exocytosis.[97] Yet, as presented here, at least two more proteins – Munc13 and synaptotagmin – establish connections between the vesicle and plasma membrane and may directly contribute to the acceleration of fusion pore opening compared to other SNARE driven fusion events of the secretory pathway. The molecular details of their interactions remain to be uncovered.

Membrane fusion is a complex process involving diverse components, the interplay of which is far from being understood to date. Genetic screens were used to identify the proteins involved in neuronal exocytosis and complemented by their biophysical and structural analysis. However, the complexity of a living cell makes observations of functional relationships made *in vivo* subject to ambiguous interpretations. Artificial model systems comprised of only few components are therefore powerful tools for elucidating the mechanisms of membrane fusion. They allow systematic manipulations that would not be possible *in vivo* e.g. for viability reasons.

Various model membranes are available to mimic synaptic vesicles and the presynaptic membrane. By selecting liposomes of different sizes and/or planar membrane setups such as planar supported lipid bilayers or pore spanning membranes membrane tension can be modulated. Also, the choice of fusion assay dictates the appropriate model membrane. Ensemble (or bulk) vesicle fusion assays, first presented by STRUCK *et al.* in 1981, have given insight to average vesicle fusion behavior by measurement of fluorescence changes.[121] Fusion can either be monitored by lipid mixing or by content mixing. For lipid mixing, two vesicle populations are prepared, and the membrane lipids are laced with lipids modified with two types of fluorophores that constitute a FRET pair (see more in detail in section 3.3.1). Upon membrane fusion, FRET efficiency is either decreased (dequenching setup where both fluorophores are in the same population and the second vesicle population is left unlabeled) or increased (quenching setup where each population is laced with one type of fluorophore). To verify true fusion in contrast to hemifusion or mixing of inner lipid leaflets caused by spontaneous lipid flip-flop, content mixing is measured. This has been achieved by filling vesicles with contents that will increase in fluorescence upon fusion. The $Tb(DPA)_3^{3-}$ chelation complex as well as self-quenching dyes like sulphorhodamine B have been exploited for this purpose.[122,123] Despite their popularity, ensemble fusion assays have some drawbacks. For example, content mixing assays notoriously suffer from leakage or rupture of the vesicles, which results in a false positive result that is hard to distinguish from fusion events. In lipid mixing assays, the observed time scales (minutes) are not suitable to zoom into the sub-millisecond velocities of Ca^{2+} triggered exocytosis and they do not report on docked states. Single vesicle fusion assays that overcome some disadvantages of ensemble assays have recently been reviewed by BRUNGER *et al.*[124] A new study by MÜHLENBROCK *et al.* even reports a setup that differentiates between rupture/leakage and fusion.[94]

Generally, two approaches in the choice of fusion mediators can be used when SNARE-induced membrane fusion is studied *in vitro*. The first approach uses native SNARE proteins (predominantly the neuronal SNARE machinery) extracted and purified after expression in a suitable organism (e.g. *E. coli*). Reconstituted into liposomes, they have been used to identify the minimal fusion machinery.[96] Yet, with the native structures being dependent on the interactions with various agents present in the neuronal cytosol i.e. for guidance of the SNARE assembly,[108] it can be difficult to isolate their effect inside of artificial setups. With regard to fusion kinetics, a major progress was made, when the ΔN49-complex was developed by POBBATI *et al.*[125] A truncated version of syntaxin-1A [183-288] lacking the H_{abc} domain was paired with a SNAP-25 mutant with all cysteines replaced by alanines so that side reactions during expression would be reduced. The 1:1 complex was stabilized by a fragment of the synaptobrevin-2 [49-96] SNARE motif which would later be displaced by the full-length synaptobrevin-2. The preorganized complex accelerated fusion to be completed within few minutes instead of hours which was ascribed to preventing the spontaneous formation of an inactive 2:1 complex in absence of the inhibitory H_{abc} domain. It has been used in different model membrane setups to e.g. to investigate fusion pathways or the effect of membrane tension on fusion kinetics. [79,126,94,63]

The second approach to in vitro studies uses artificial fusogens, designed to mimic different aspects of the SNAREs. Relatively easy synthetic access and high control over the connectivity and geometry of the SNARE analogues provides the opportunity for systematic variation to study different facets of membrane fusion on a molecular level. SNARE analogues – similar to their native archetype – typically consist of three parts: recognition units, flexible linkers, and membrane anchors. The artificial fusogens presented in the following paragraphs (grouped by interactions exploited in the recognition units) have been used for detailed investigations of the roles that each module plays in promoting membrane fusion. The late 2000s and early 2010s marked a boom phase for targeted membrane fusion. Multiple new artificial SNARE mimetics have been presented within a period of approximately 4 years.[17,18]

Small molecule recognition

Diverse small molecule interactions have been successfully used to efficiently mediate membrane fusion. Specific recognition between boronic acid (linked by a PEG spacer to a stearic acid anchor) and *cis*-diols as found in the sugar-like head group of phosphatidylinositol has been exploited by KASHIWADA *et al.* to promote liposome/liposome fusion.[19] Complemented by pH-responsive coiled coil peptides, boronic acid/cis-diol recognition could be used to generate temporal control over liposome fusion, triggered by pH change.[127] Hydrogen bonding between cyanuric acid and melamine, attached to liposome membranes via a lipid anchor was another reported model system showing fusogenic properties.[128] Within this group, also a recent fusogen based on strain-promoted azide-alkyne cycloaddition may be noted.[129] The covalent connection introduced by click-reaction was reported to promote fusion between liposomes.

The winning arguments for these kinds of model fusogens are their simplicity and biocompatibility, making them an interesting target for drug delivery. However, the SNARE-mimicking aspects are rather limited as the directionality and zippering of SNARE complex assembly cannot be addressed.

Nucleobase-pairing

To mimic zipper-like assembly of the SNARE complex, several model fusogens with interactions between oligomers displaying nucleobases have been developed. STENGEL *et al.* presented a model system in 2007 that places complementary DNA strands in separate liposome populations, anchoring them with cholesterol (see Figure 1.6 A).[20] A zipper-like orientation was achieved by connecting the lipid at the 5' and 3' end, respectively. Efficient lipid mixing was observed, but only ~30% could be attributed to inner lipid mixing, i.e. full fusion. In a follow-up study, the effects of varying the length of the duplex and introduction of either a mono-anchor (composed of PEG$_4$ and cholesterol) or a double anchor (two chains of PEG$_4$-cholesterol per single strand) were probed.[130] Regarding the duplex length, between the used 12-, 27- and 42mers the 27mer showed the strongest fusogenicity, further elongation did not improve fusion. A double anchor, in this case, proved essential to full fusion while single anchors mostly produced outer leaflet mixing. This effect was attributed to an increased stiffness of the double linker, expressed in better force translation to the lipids. A similar construct was used to induce targeted fusion between liposomes and SLBs.[131] Liposomes docked to the SLBs by DNA duplex formation could be selectively promoted to fusion by addition of Ca^{2+}. The divalent cation is known to interact with phospholipid headgroups, perturbing the organization of the bilayer and can by itself induce membrane fusion in appropriate setups.[132] DNA-base pairing was also used by CHAN *et al.* to show the importance of duplex orientation.[133] In ensemble fusion assays complementary strands both anchored to the liposomes via the 5' end (resulting in non-zipper-like dimerization) did not produce membrane fusion contrary to the corresponding zipper-like setup.

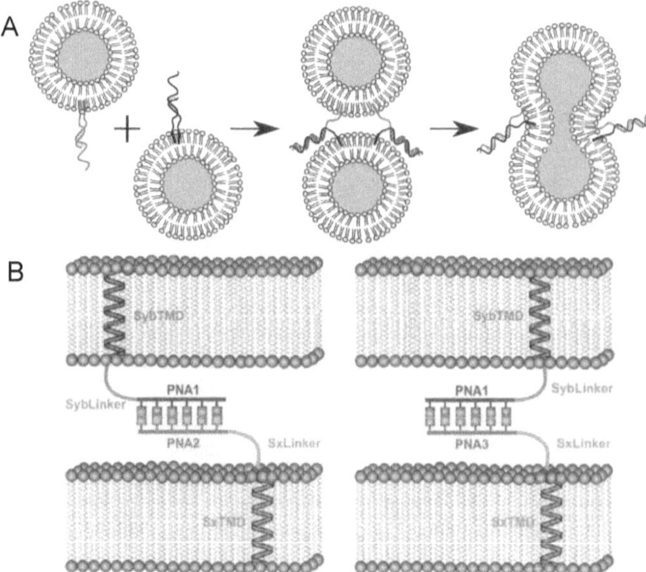

Figure 1.6: A) DNA-based model system presented by STENGEL et al.[130] B) PNA SNARE hybrids by LYGINA et al. with duplex orientation and stability easily modulated by the PNA sequence.[134]

A different backbone topology was exploited with the development of a model system based on peptide nucleic acid (PNA) recognition by LYGINA *et al.*[134] These model fusogens differed from other known fusogens by the choice of anchor, attaching the highly thermally stable recognition units to linker regions and TMDs of native neuronal SNAREs syntaxin-1A and synaptobrevin-2 as shown in Figure 1.6 B. A characteristic feature of the PNA duplexes is that orientation of duplex assembly can be readily defined by the nucleobase sequence which was also used for the described PNA/SNARE hybrid. In ensemble fusion assays, parallel orientations could produce a higher rate of membrane fusion than antiparallel orientation. Additionally, the important role of TMD sequences could be highlighted by the observation that the use of identical TMDs or truncated TMDs significantly reduced fusion efficiency. TMDs were further investigated by WEHLAND *et al.*, reporting a significant enhancement in fusogenicity when *C*-terminal amino acids (usually anionic because of the carboxy-terminus) were exchanged for neutral net charge amino acids as lysine and reduction in fusogenicity when dianionic amino acids as glutamic acid were used.[135] HUBRICH *et al.* later demonstrated the potency of the model system, showing that as few as 5 (aeg) PNA monomers were sufficient to generate efficient membrane fusion.[136] Another variation of the PNA/SNARE hybrid was presented by SADEK *et al.* with β-peptide/β-PNA as recognition units. The additional methylene units in the peptidic backbone produce a stable and highly rigid 14-helix, with a very predictable sequence-to-structure relation. Every fourth residue of the recognition unit was decorated with a nucleobase, yielding a Watson-Crick interaction site. This setup was used to study the distance dependence of antiparallel duplex assembly. With an antiparallel 4-basepair recognition unit, full fusion could be verified. However, when a rigid β-peptide spacer was placed at the *C*-terminal end of the recognition units, only hemifusion was achieved.

Coiled-coil interactions

The first model fusogen based on interactions of coiled coil forming peptides E3 ((EIAALEK)₃) and K3 ((KIAALKE)₃) was presented by MARSDEN *et al.* in 2009.[137] They exploited the heterospecific duplex designed by LITOWSKI *et al.* to obtain targeted recognition of liposomes.[22] The peptides were incorporated into liposomes by a lipid anchor, DOPE, bridged by a short PEG₁₂ (spacer LPE and LPK, see Figure 1.7 A). It was proposed that parallel heterodimer assembly would dock the liposomes together, mimicking the zippering of SNARE complexes as shown in Figure 1.7 B. Efficient lipid mixing and content mixing could be demonstrated in ensemble fusion assays. Observations from cryo electron microscopy and optical microscopy of 100 nm and 1 μm liposomes which both demonstrated signs of fusion led to the conclusion that curvature stress determined by liposome size was not the main driving force for fusion. In follow-up studies, the mechanism by which membrane-bound E3 and K3 mediate fusion was investigated more in depth by CD and IR spectroscopy and membrane compression analyses.[138,139] In addition to the vesicle docking by duplex formation proposed earlier, it was found that K3 more than E3 interact strongly with the external lipid monolayer, incorporating as α-helical monomers parallel to the lipid plane. This interaction mode was suggested to facilitate membrane fusion by induction of curvature and disruption of lipid-lipid interactions, lowering the energy barrier to fusion stalk formation.

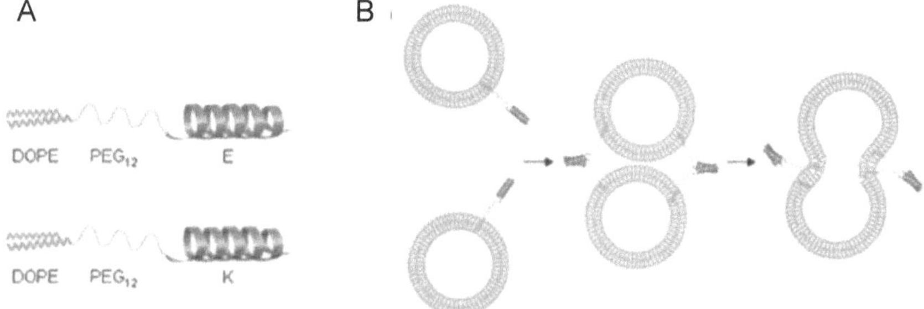

Figure 1.7: A) Illustration of lipopeptides LPE and LPK developed by Marsden et al. B) Proposed course of liposome fusion promoted promoted by LPE and LPK. Modified in accordance with MARSDEN et al.[137]

The impact of different modifications to the lipopeptides have been tested in several studies. Variations of anchoring lipid and spacer lengths showed asymmetric effects on E3 and K3 lipopeptides.[140] While in K3 bound to cholesterol a decreased linker length (PEG₄ or PEG₈) led to stronger interactions with the membrane, the same tendency could not be found in DOPE bound K3. In E3 lipopeptides, increased linker length (PEG₁₆) proved beneficiary for membrane fusion. Generally, cholesterol-bound lipopeptides were more fusogenic than their DOPE counterparts. The significance of orientation of the coiled coil assembly has also been a subject of investigation. Antiparallel assembly has been pursued by two approaches, a – by attaching linker and lipid to the N-terminus of one of the peptides which still yields a parallel coiled coil complex but with non-zipper-like peptide orientation – and b – by sequence inversion, leading to true antiparallel duplex orientation.[141,142] On the other hand, a publication by PÄHLER *et al.* reported contradictory results[143] Constructs of E3, K3 and inverted sequences iE3 and iK3 were extended by a short peptidic linker containing a cysteine at the *C*-terminus were investigated by CD spectroscopy as monomers and as combinations forming parallel and antiparallel coiled coils. Furthermore, the peptides were bound *in situ* to separate populations of liposomes (exploiting cysteine/maleimide reactivity) and parallel and antiparallel combinations were tested regarding fusogenicity in ensemble lipid mixing assays. While in all heterodimeric combinations dissociation constants calculated from CD titrations were similar (between 2 μM and 7 μM), the propensity to induce lipid mixing was drastically lower in antiparallel complexes than in parallel combinations.

The E3/K3 lipopeptides have also found practical applications directed towards drug delivery. Delivery of the cytotoxic drug doxorubicin encapsulated into liposomes could be achieved by liposome/cell membrane fusion *in vivo*.[144] Furthermore, temporal control over liposome/liposome fusion could be achieved employing a photolabile protection strategy.[145]

In 2011, another SNARE mimetic adopting E3/K3 recognition has been presented.[26] MEYENBERG *et al.* used the native SNARE TMDs and linker regions of synaptic syntaxin-1A and synaptobrevin-2 and replaced the SNARE motives with the coiled coil forming peptides yielding the SNARE mimetic pair E3Syb and K3Sx as shown in Figure 1.8 A and B. Thus, the complexity and length of the tetrameric SNARE complex was reduced to the ~20 amino acid long heterodimeric coiled coil. The continuous topology of fully peptidic SNARE analog was proposed to more closely reflect the buildup of SNARE proteins compared to other model fusogens, allowing conclusions more relevant to membrane fusion *in vivo*. The constructs have some drawbacks, for example, the attachment of the recognition units has been chosen in a way that does not allow the continuation of interactions between linker and TMD amino acids as they would be found in the native SNAREs. Also, considering the findings made in the KROS group for K3 lipopeptides,[138–140] the K3Sx peptides likely exhibit membrane destabilizing properties that are not found in the native system. Nevertheless, the fusion pair has been found to heterospecifically promote membrane fusion in ensemble fusion assays with kinetics that are comparable to the ΔN49 complex and has since been used to study the roles of linker amino acids and TMDs for membrane fusion.[26,146,147] In this book, the peptidic SNARE analog will be used to obtain temporal control over membrane fusion applying appropriate photolabile protection strategies.

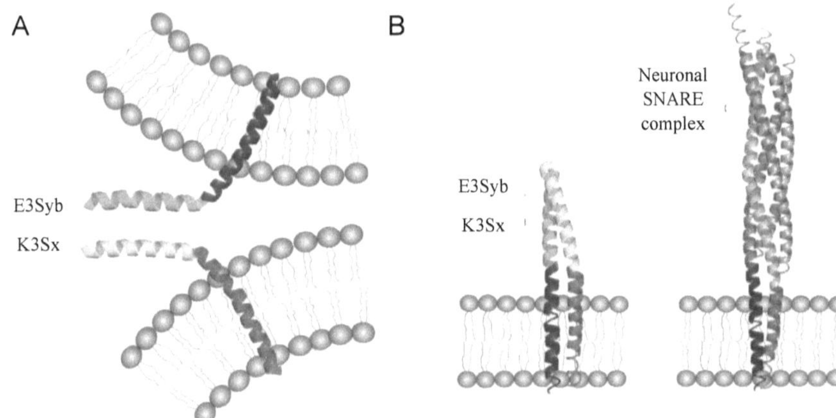

Figure 1.8: Fully peptidic SNARE mimetic developed by KARSTEN MEYENBERG. A) Illustration of the proposed interactions of model peptides reconstituted into liposomes. B) Comparison of the structures of the putative cis-E3Syb·K3Sx complex and the neuronal SNARE complex.[98] Figure modified in accordance with MEYENBERG.[146]

All proteins produced in an organism are closely monitored by the immune system for signs of disease. While most proteins are catabolized back to amino acids in proteasomes on a regular basis, a small portion of semi-processed oligopeptides is transported by the transporter associated with antigen processing (TAP) to the endoplasmic reticulum (ER).[148,149] There, they are cut down to peptides of 8-11 amino acids[32] and loaded onto major histocompatibility complex I (MHC-I) to be presented on the cell surface and inspected by CD8[+] T-cells. CD8[+] T-cells are trained to be tolerant to autologous proteins, but when a cell expresses mutated sequences or is infected by a virus, antigenic peptides are identified, and the cell is eliminated.[150] CD8[+] T-cells do not patrol the organism themselves before being activated. There is an immense number of different CD8[+] T-cells, specific to every antigenic amino acid combination, but of every type only a limited number is kept in stock, distributed in secondary lymphoid tissues and concentrated in lymph nodes.[151] It is the task of dendritic cells to collect cell samples from peripheral tissues by endocytosis and phagocytosis,[152] transport them to the lymph nodes and present them on MHC-I to the naïve CD8[+] T-cells.[33] This process is called cross-presentation. When an antigen is recognized by a CD8[+] T-cell it is activated which dominoes into a series of adaptive immune responses to eliminate the diseased cells. After uptake by the dendritic cell, different mechanisms are known by which exogenous proteins are processed and loaded on MHC-I for cross-presentation. The phagosome-to-cytosol and the vacuolar pathway are discussed below. Similar to the CD8[+] T-cells, there are many different MHC-I variants with slightly different binding grooves. Polymorphic human leucocyte antigen-A and B code for the antigen presenting molecules and more than 8500 individual haplotypes have been be identified so far.[153] The different haplotypes bring along preferences for polarity and charge of the individual amino acid positions building up the presented epitope.[154]

1.2.1 Phagosome-to-cytosol (P2C) pathway

Antigens from exogenous proteins taken up inside phagosomes can leave the endocytic compartment to be processed by proteasomes in the cytosol (Figure 1.9). Once broken up by the proteasome, peptides are transferred inside the ER by TAP or back into the phagosome to be loaded on MHC-I by peptide loading complex (PLC). Several steps of the P2C pathway have been suggested to involve the localization of components in phagosomes that are usually found in the ER. Among others, MHC-I, TAP and tapasin, all part of the PLC, have been verified in phagosomes by mass spectrometry, however their functionality in the phagosome is under debate.[36] Some components may be transferred from the ER by a process involving Sec22b, a vesicle trafficking protein from the SNARE family.[155] This has been supported by silencing Sec22b, which leads to reduced cross-presentation. With this finding it has been discussed if ER-associated degradation (ERAD) mechanisms might play a role in transporting antigen out of the phagosome to the cytosol.[156] Additionally, silencing an enzyme involved in ERAD –

ubiquitin-conjugating enzyme which marks proteins for degradation – was demonstrated to inhibit cross-presentation.[157] This step may be complemented by occasional rupture of the phagosome membrane due to radical oxygen species releasing yet undigested and functional exogenous proteins to the cytosol.[35,158] Once transported back inside the eligible organelles (phagosome or ER) by TAP, proteasome-degraded oligopeptides may require further trimming to fit the receptor of MHC-I. This task is fulfilled by insulin-regulated and ER aminopeptidases (IRAP and ERAP).[157] The origin of MHC-I in proteasomes can be twofold, being recycled from the cell surface or newly synthesized from the ER. However, for the P2C pathway transport from the ER seems to contribute the bigger portion.[157]

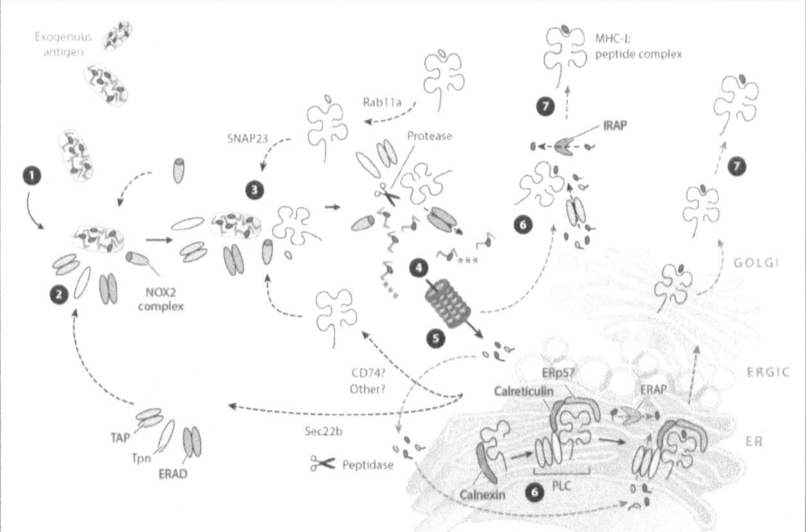

Figure 1.9: Cross-presentation by the phagosome-to-cytosol pathway. Foreign material is taken up by the dendritic cell inside of an enclosed compartment, the phagosome. Predegraded or intact protein is transferred to the cytosol where it is broken up by proteases. TAP delivers the peptides to phagosomes or ER, where they may require a final trimming before being loaded on MHC-I. [34]

1.2.2 Vacuolar pathway

Rather than leaving the phagosome, foreign proteins can also be converted to MHC-I appropriate bites by protein catabolism inside the endocytic compartment (Figure 1.10). This pathway is neither proteasome nor TAP dependent, which makes proteasome and TAP inhibition good tools for identifying how a specific antigen is generated. Instead, proteases (especially Cathepsin S) and proton pumps are delivered to the phagosome by fusion with lysosomes.[38] Acidification of the compartment activates the proteases and thereby facilitates proteolysis. Final trimming of the peptides is accomplished by IRAP if required before loading onto MHC-I. For the vacuolar pathway, recycled MHC-I from the cell surface was found to be the major source of MHC-I in the phagosome/lysosome.

Interestingly, MHC-II molecules are loaded with antigens to present to CD4$^+$ T cells by similar mechanisms.[159] In a recent study functional proteasomes were found to be active in endocytic organelles disputing the fact that one can distinguish the vacuolar pathway from the P2C pathway by proteasome inhibition.[160]

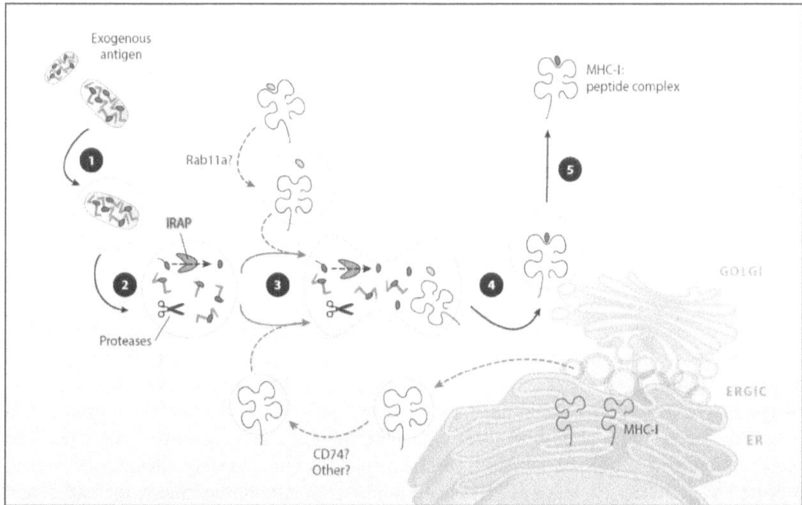

Figure 1.10: Cross-presentation by the vacuolar pathway. Antigen filled phagosomes fuse with lysosomes which deliver proteases. Proteolysis requires the compartment to be acidified by proton pumps. MHC I may be recycled from the cell surface before being reloaded with antigen and transported to the plasma membrane to present their cargo to CD8$^+$ T cells. [34]

1.2.3   Cross-presentation of membrane buried epitopes

The preference of MHC-I for binding a specific epitope is predefined by their HLA-A or B haplotype, the multitude of which can be grouped in five HLA-A and seven HLA-B super types.[154] Bioinformatic tools have been found to be accurate predictors for the epitopes which are bound by these super types.[161,162] In a study by BIANCHI *et al*, epitopes predicted to be presented by the different haplotypes were correlated with typical transmembrane helix (TMH) compositions and a major overlap was found (Figure 1.11).[39] Especially in HLA-A02, the most abundant haplotype among the Caucasian population, the overlap was striking.

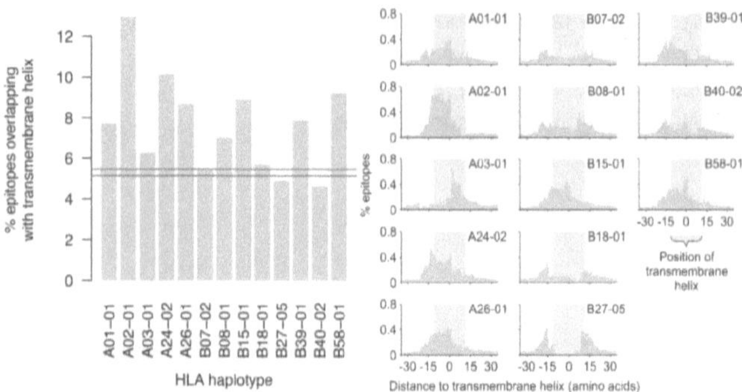

Figure 1.11: Left) Predicted epitopes derived from the human proteome bound by all HLA-A and HLA-B super types that overlap with predicted TMHs by minimum one residue. Red bars mark the threshold for random correlation. Right) Percentage of epitope middle positions plotted against their distance (aa residues) to the nearest TMH central position.[39]

Analysis of epitopes naturally presented in B lymphoblastoid cell lines partly supports this connection, showing that about 1% of the presented epitopes can be predicted to originate from TMHs.[163] To be loaded on MHC I, antigens of both endogenous and exogenous origin would need to be extracted from the lipid membrane they are embedded in and delivered to proteolytic entities e.g. proteasome in the cytosol or lysosomes. The presentation and cross-presentation mechanisms described above have not yet been reported to include such steps. TMHs need to meet specific requirements for length and amino acid composition to insert in the right membranes with the right orientation and are therefore often highly conserved.[164] Cross-presentation of membrane buried epitopes might be a valuable tool to keep otherwise rapidly mutating pathogens in check.

2 Synthesis and purification of difficult peptides

About 70% of the peptides synthesized in this work can be considered as "difficult peptides". The term has been coined in the late 1980s to summarize the phenomenon of sequence dependent synthesis failure in solid phase peptide synthesis (SPPS).[165] In any given sequence, the reaction efficiency is never 100% for each of the reaction steps. Formation of side products with only one or two missing chain members with respect to the full-length sequence and accumulation of errors in long sequences is therefore routinely observed. However, repetition of amino acids or the abundance of β-branched amino acids threonine, isoleucine and valine have been shown to aggravate synthetic problems by inter-chain association of the growing oligomer or interactions with the matrix of the solid support.[165,166] Nowadays, peptides that are hard to purify are also considered representatives of this group.[167] Poor solubility in common solvents such as aqueous buffers or mixtures of water and MeOH or MeCN can originate in different properties of the peptides (Table 2.1). High content of hydrophobic amino acids is typical in transmembrane peptides such as TM9_pra8 (**1**). Self-assembling behavior can be based on the amphiphilic character of a peptide like in E3Syb (**2**) or the abundance of hydrogen bond forming amino acids glutamine, serine and threonine like in NY-ESO1_long_pra5 (**3**).

Table 2.1: Selected difficult peptides representative of the peptides synthesized for this work.

H-AAAWPFVLLCLQQLSLLMWIT{pra}CFLWAAA-OH TM9_pra8 **1**	Abundance of neutral and hydrophobic amino acids dominate behavior in solution.
H-G(EIAALEK)₃RKYWWKNLKMMIILGVICAIILIIIIVYFST-OH E3Syb **2**	Charged domain adjacent to hydrophobic domain presents a potential for surfactant-like self-organization in solution.
H-LQQLSLLM{pra}ITQCFL-OH NY-ESO1_long_pra5 **3**	Strong hydrogen bonds between residues promote self-organization and aggregation in solution.

Peptides used in this work were mostly synthesized by automated SPPS supported by microwave irradiation with the Fmoc protection strategy (Scheme 2.1). The used protocols were adapted from methods published by the manufacturer of the peptide synthesizer.[168,169] Related protocols have previously been used to produce peptides of high hydrophobic content and length.[147,170] As solid support either low load Wang resins, preloaded with the first amino acid, or H-Rink-Amide Chemmatrix® resin were used. The Wang resins provided carboxy *C*-termini and were used with low loading density (0.27-0.32 mmol/g) to reduce aggregation of the peptides on resin. Amide *C*-termini were only used on medium length soluble peptides and thus a medium loading density was applied (0.5 mmol/g). Both resin types are considered high swelling in *N,N*-dimethylformamide (DMF) and contribute to the success of peptide synthesis.

In automated SPPS, deprotection of *N*-terminal Fmoc was always accomplished with 20% piperidine in DMF within 90 s at 90 °C. To choose the method for building block coupling the length of the peptide was considered and whether the peptide could subsequently be purified by chromatographic methods. For peptides that would be soluble in water or a mixture of water and MeCN or MeOH, single coupling was performed. The commercially available building blocks (5 eq.) were added to the resin together with activator DIC (5 eq.), racemization preventing additive Oxyma Pure (5 eq.) and DIPEA (1 eq.) which restricted the formation of Oxyma Pure derived precipitate. Coupling was completed within 120 s by MW irradiation at 90 °C. For the synthesis of peptides that would be hard to purify or that would have to be used crude, the CarboMAX™ method, developed by CEM, was employed.[169] The use of excess carbodiimide with respect to the amino acid was shown to be beneficial in the synthesis of difficult peptides. Additionally, with the CarboMAX™ method the amino acids were coupled twice.

It was reported that by using excess carbodiimide in absence of base, otherwise observable epimerization of susceptible amino acids like cysteine could be suppressed even at high temperatures.[169] We therefore tried to conduct the synthesis of E3Syb (**2**) without the addition of DIPEA and couple all amino acids at 90 °C. Without base, the synthesis reproducibly failed after position 17 or 18 of the peptide due to the buildup of precipitate that hampered free flow of reactants and solvent. The precipitate could not be removed even by multiple washing steps with DMF, only washing with aqueous solvent could free the filter of the reaction vessel. With addition of DIPEA, this precipitate was not observed so the tradeoff was made to couple cysteine at 50 C for prolonged periods (10 min vs 2 min).

Synthesized building blocks were chosen to be coupled manually for two reasons. Firstly, the coupling behavior in microwave assisted SPPS was mostly unknown, so the reaction would need to be followed by Kaiser-test and access to resin samples is more convenient in manual synthesis than in the pressurized peptide synthesizer. Secondly, excess of building blocks could be more easily collected to be recycled. In a related bachelor book, it was observed that coupling conditions with DIC/Oxyma (5 eq. each, 5 eq. aa, 90°C) that worked well in automated synthesis could not be directly translated to manual coupling in an open reaction vessel.[171] Instead, manual coupling was achieved by activation with HATU/HOAt (5 eq./4.5 eq) and DIPEA (10 eq.) as activator base and microwave heating to 75 C. Removal of Fmoc in manual cycles was adapted from a previously established two-step deprotection with 20% piperidine.[146]

Global deprotection and cleavage of peptides from the dried resin was achieved with TFA/TIS/H$_2$O (95:2.5:2.5, *v/v*) or TFA/TIS/EDT/H$_2$O (94:2.5:2.5:1, *v/v*) where EDT was only used to prevent disulfide bridge formation when cysteine was present. Some peptides were modified after linear synthesis of the amino acid sequence and before acidic cleavage. This includes acetylation to remove charge from the *N*-terminus, *N*-terminal fluorescent labeling and ring-closing metathesis.

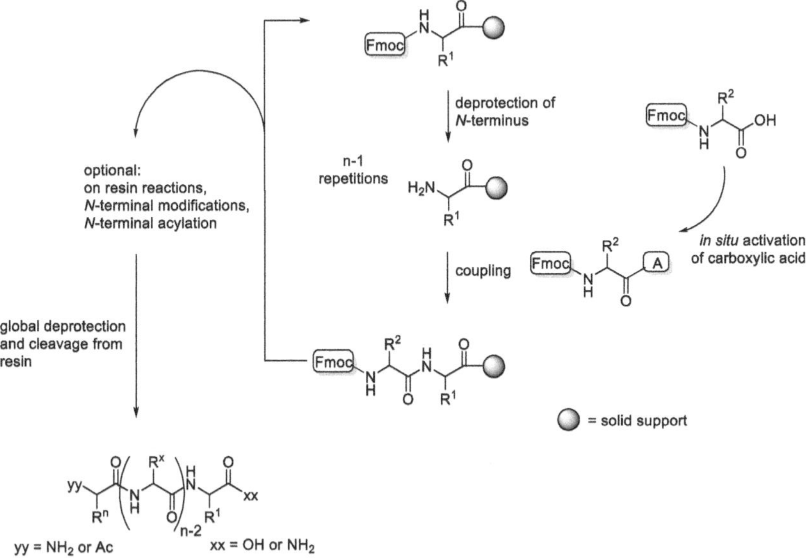

Scheme 2.1: Simplified representation of solid phase peptide synthesis.

The standard method for post SPPS purification is high performance liquid chromatography on reverse phase (RP-HPLC) with a gradient of aqueous buffer mixed with acetonitrile or methanol, often adjusted to acidic pH with 0.1% TFA.[172] However, standard purification protocols can often not be employed for aggregation-prone peptides due to poor solubility. Removable backbone modification with polyArg solubility tags has been reported to improve peptide behavior during HPLC[173] and while synthetic effort is acceptable in a single peptide, it can accumulate to be inconvenient when many different peptides must be purified. Instead, we tested trifluoroethanol (TFE) containing eluent mixtures developed for transmembrane peptides by HARA *et al.* with our peptides dissolved in TFE.[12] While the peptide solutions appeared completely homogenized, purification was not successful. Only a small portion of injected crude material could be recovered after elution during the washing step with pure organic solvent. Further analysis revealed that no separation could be achieved. We concluded, that incomplete monomerization of putative peptide aggregates were the cause of failure to purify.

Proper solubilization was achieved with 1,1,1,3,3,3-hexafluoroisopropanol (HFIP). The strong H-bond donor is known for inducing and stabilizing α-helical structure and thus separating inter-chain associations.[174] HFIP was not the first choice as solvent for HPLC samples because of anticipated preelution of samples and because it is not miscible with H_2O/organic solvent mixtures at all ratios. Still, dilution of peptide/HFIP solutions with H_2O right before injection proved beneficial for retention of crude peptides on C18 columns and the peptides could be eluted and separated with gradients of H_2O and MeOH or H_2O and MeOH/1-PrOH 4:1 (Figure 2.1).

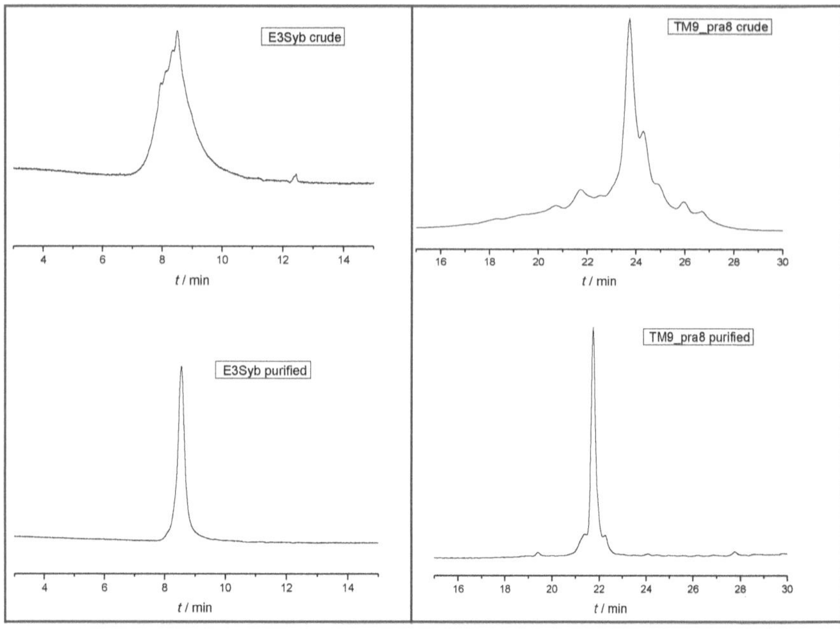

Figure 2.1: Left) UPLC chromatograms of E3Syb prior to purification and after HPLC. Elution from column 1 was achieved by a gradient of 70 to 100% B (solvent system II) with a flow of 0.4 mL/min in 8 min and subsequent isocratic elution with 100% B. Right) HPLC chromatograms of crude TM9_pra8 (column 5, 70 to 100% B in 30min, flow 10 mL/min, solvent system VI, 50 °C) and purified TM9_pra8 (column 2, 70 to 100% B in 30min, flow 1 mL/min, solvent system VI, 50 °C).

Using HFIP for sample preparation, SNARE analoga like E3Syb (**2**) and artificial transmembrane peptides with incorporated antigens like TM9_pra8 (**1**) could be routinely purified on HPLC using standard eluents H_2O and MeOH and modifying hydrophobicity with varying contents of 1-PrOH if needed. However, the method has its limitations. 15 amino acid peptide NY-ESO1_long_pra5 (**3**) and various derivatives that were synthesized as presumably soluble antigens could not be fully homogenized in HFIP. Mixtures of **3** with the fluorinated solvent appeared clear but could not pass through a syringe filter (45 µm pore size). We attempted to solubilize the peptides with a protocol developed by BURRA *et al.* for polyGln peptides.[175] For this, the peptides were pretreated with TFA/HFIP to monomerize aggregates and after solvent removal by gentle N_2 stream, formic acid was added. The solution was to be diluted to 20% formic acid by addition of H_2O, however, any addition of H_2O caused immediate visible precipitation. Only minor side products of SPPS could be observed by ESI mass spectrometry (Figure 2.2), so this peptide class was used crude after thorough washing with Et_2O.

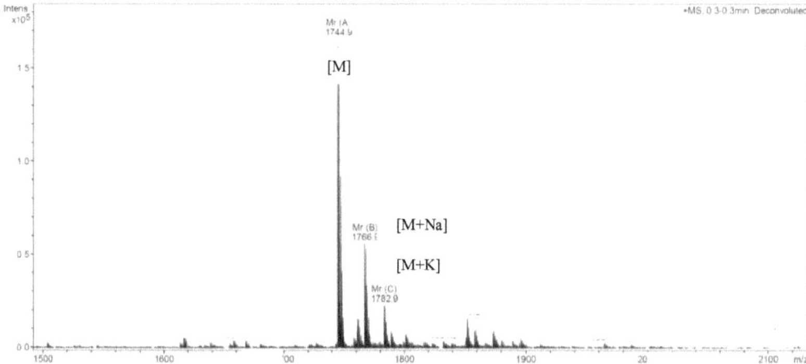

Figure 2.2: Deconvoluted ESI MS spectrum of crude NY-ESO1_long_pra5 (3).

3 Modulating E3Syb/K3Sx fusion with common photocleavable protecting groups

In this study, SNARE protein analogs developed from neuronal synaptobrevin-2 and syntaxin-1A were used. The tetrameric recognition motif of the native fusogens was replaced with coiled coil forming peptides E3 and K3 to form the model fusion pair E3Syb/K3Sx. In bulk fusion experiments, E3Syb/K3Sx developed by MEYENBERG *et al.* had been shown to induce full fusion of large unilamellar vesicles[26] and was now used as a base system to control membrane fusion by a photolabile protection strategy. In previous studies mutations in the minimal fusion machinery were used to control giant unilamellar vesicles in docking and hemifusion stages.[79,176] In absence of a suitable trigger, no recovery of fusogenicity was attempted. Temporal control of membrane fusion has been achieved by KONG *et al.* using photocleavable PEGylation and artificial fusogens with coiled coil forming recognition units.[145] The steric shielding in this approach completely suppressed interactions between recognition units and vesicles and could not be used to study intermediate states of fusion.

It was aimed to develop a protection strategy that would suppress lipid mixing but still allow docking of vesicles. Arresting the model system in a preorganized state would provide an immediate response to a photo-trigger, mimicking primed SNARE complexes responding to calcium influx. Later, if a hemifusion state can be stabilized, this would allow to test whether the intermediate can be converted to full fusion (see section 1.1.1). The use of the native TMDs and linkers of neuronal SNAREs then allows studying their influence on transitions between fusion stages. To accomplish this kind of control, the recognition units of E3Syb/K3Sx were targeted as depicted in Figure 3.1 A and C. Two *N*-terminal heptads of the coiled coils were to remain unchanged and, in the membrane-proximal heptads, coiled coil interactions were to be disturbed with PPGs.

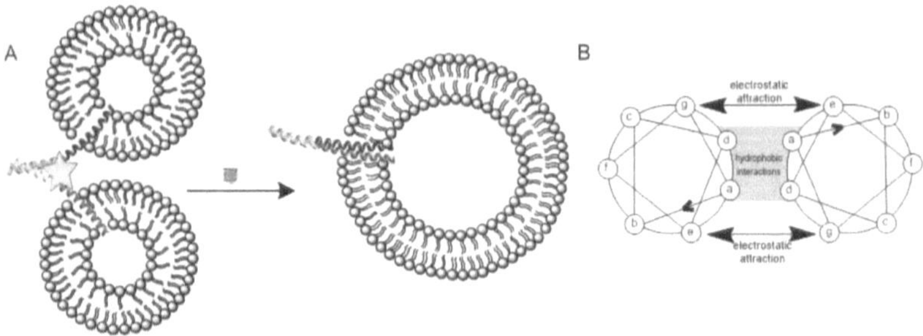

Figure 3.1: A) Illustration of the pursued protection principle and control over vesicle fusion. Caged E3Syb (E3 green, linker and TMD of synaptobrevin-2 blue) and K3Sx (K3 yellow, linker and TMD of syntaxin-1a red) reconstituted into vesicles interact via the N-terminal region of the coiled coils to allow docking of vesicles. Interactions of the membrane-

proximal heptads are inhibited by PPGs (up to four individual PPGs, represented by the yellow star) so lipid mixing membrane fusion is suppressed. Upon release of the PPGs by UV irradiation activity of the preorganized fusion pair is restored and leads to vesicle fusion. B) Helical wheel representation of interactions in dimeric coiled coils. C) Peptide sequences of E3Syb and K3Sx (recognition units with pale green and pale yellow background, linker regions and TMDs with pale blue and pale red backgrounds. Electrostatic interactions of E3 and K3 are depicted as dashed lines. Amino acids, that can be reasonably caged in the membrane-proximal heptad are highlighted in yellow. The assignment of the heptad register can be seen below the recognition unit of K3Sx.

When trying to reversibly inhibit coiled coil interactions in E3 and K3,[177] two main contributions have to be considered: charged amino acids in positions e and g (Figure 3.1 B and C) exhibit electrostatic attraction when paired with the complementary binding partner and electrostatic repulsion when approached by a peptide of the same kind, thus guaranteeing heterospecificity. Hydrophobic interactions between isoleucine and leucine (a and d) stabilize the duplex. Within the amino acid composition of E3 and K3 only lysine and glutamic acid are synthetically approachable to be caged by PPGs. Therefore, several Fmoc-L-Lys(PPG)-OH and Fmoc-L-Glu(PPG)-OH derivatives were synthesized and tested for the desired application.

3.1 Photocleavably protected amino acid building blocks

3.1.1 6-Nitroveratryl-based caged amino acids

The nitroveratryl (NV) group is a derivative of the *o*-nitrobenzyl group first introduced as a PPG in 1970.[178] Methoxy groups in 4- and 5- positions induce a bathochromic shift of the absorption maximum enabling cleavage at ~360 nm. However, shorter wavelengths are more effective.[179] Photolysis occurs under the mechanism shown in Scheme 3.1Scheme 2.1.[180]

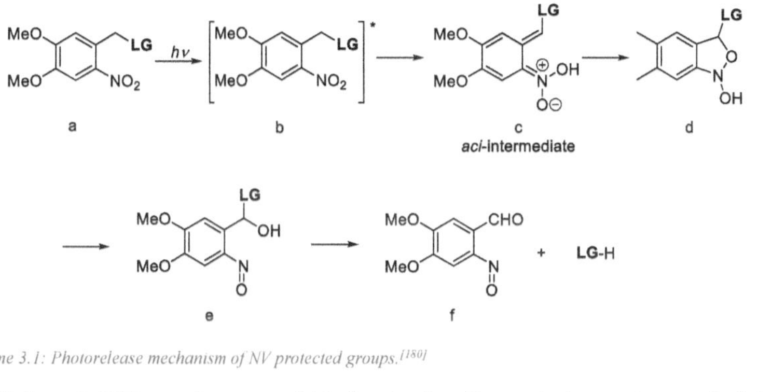

Scheme 3.1: Photorelease mechanism of NV protected groups.[180]

Irradiation of a NV-caged compound (a) elevates the nitroveratryl group to an exited state (b). Through intramolecular proton shift the *aci*-intermediate (c) is formed which converts to the isoxazolidin-1-ol intermediate (d) by irreversible cyclization. Ring opening product hemiacetal (e) is hydrolyzed in the rate-limiting step releasing the leaving group. Byproduct of the photolysis is *o*-nitroso veratraldehyde (f), a potential toxin to surrounding bioprocesses. Despite poor quantum yield ($\Phi = 0.006$, 365 nm),[179] NVOC is one of the most popular PPGs both for applications in solution and on solid support. Due to its straightforward and cost-effective synthesis it was tested in this project.

Fmoc-L-Lys(NVOC)-OH (**4**) was synthesized as part of a related master book in a two-step procedure (Scheme 3.2).[181] 6-nitroveratryl alcohol (**5**) was reacted with 4-nitrophenyl chloroformate (**6**) to produce activated anhydride (**7**). Nucleophilic attack from the ε-amino group of Fmoc-L-Lys-OH at the carbonate gave Fmoc-L-Lys(NVOC)-OH (**4**) with an overall yield of 42%.

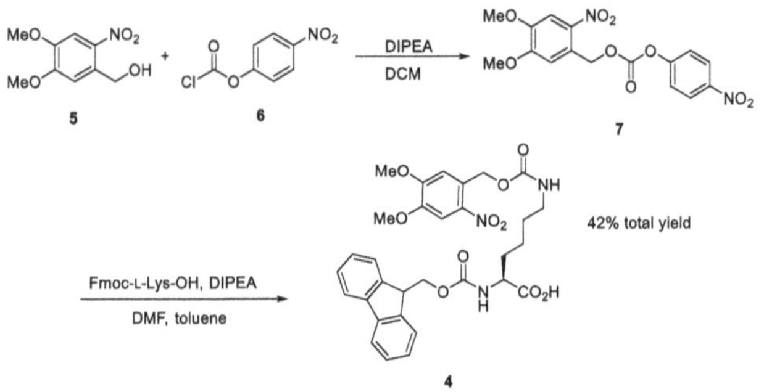

Scheme 3.2: Synthesis overview of Fmoc-L-Lys(NVOC)-OH.[181]

Deprotection rate was evaluated with the available cleavage setup by method a) (see section 7.1.5) in methanol (3.4 mM) and followed by analytical HPLC (see the appendix, Figure-A 1). Solvent and concentration were chosen diverging from the final application conditions (HEPES buffer, ~0.7 µM) to effectively dissolve the amino acid and clearly observe it in analytical HPLC. After 10 min of irradiation only a trace of the caged amino acid could be detected.

Despite being a common protecting group for carboxylic acids, NV was disqualified from being used on glutamic acid for not withstanding reaction conditions of solid phase peptide synthesis.[182] Alternatively, the closely related 3-(4,5-dimethoxy-2-nitrophenyl)-2-butyl (DMNPB) group was chosen to protect the glutamic acid residues.[183,184] It has been developed to overcome some of the drawbacks of NV by having a higher quantum yield ($\Phi = 0.26$, 365 nm) and a less toxic photolysis byproduct.[184] In literature, DMNPB protected carboxylic acids (g) are reported to traverse *aci*-nitro intermediate (h) and finally release free carboxylic acid and nitrostyrene derivative (i) (Scheme 3.3).[184] However, uncaging tests in methanol (3.4 mM) revealed multiple product peaks in analytical HPLC analysis (see the appendix, Figure-A 2), suggesting a more diverse byproduct composition. On the other hand, the caged amino acid was completely consumed after 5 min of irradiation, confirming a more effective photorelease.

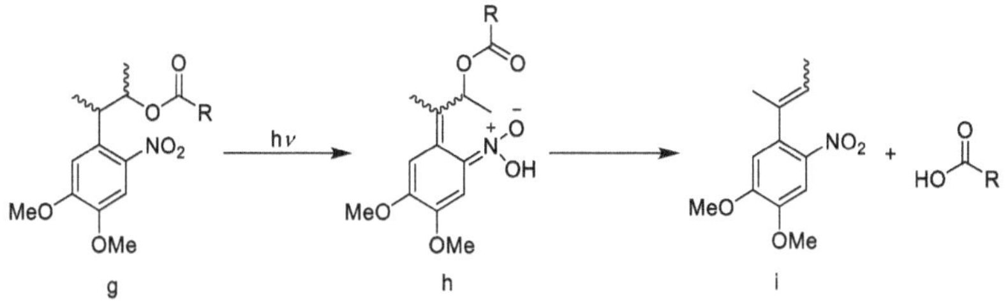

Scheme 3.3: Photorelease of a DMNPB protected carboxylic acid.[184]

Synthesis of the SPPS-ready Fmoc-L-Glu(DMNPB)-OH (**8**) was adapted from WIRKNER *et al.*[183] with minor changes and performed by HOA NAM NGUYEN for his bachelor book (Scheme 3.4).[185]

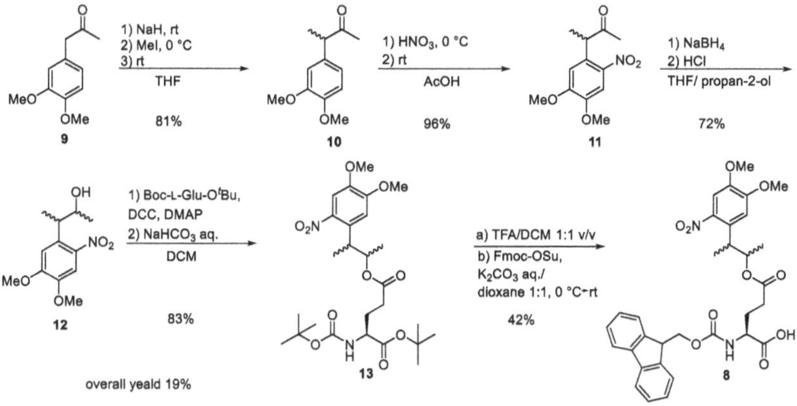

Scheme 3.4: Synthesis overview of Fmoc-L-Glu(DMNPB)-OH.[185]

First, commercially available 3,4-dimethoxy-phenylacetone (**9**) was methylated with NaH and methyl iodide to give ketone (**10**) as a racemic mixture. Postponing reduction of the carbonyl until after nitration allowed the use of aggressive nitrating acid to give compound **11** in high yields. Reduction to **12** could then be chemoselectively achieved with NaBH4. STEGLICH esterification connected the protecting group to the glutamate side chain to produce **13** and acidic hydrolysis of ᵗBu and Boc followed by Fmoc protection of the amine yielded Fmoc-L-Glu(DMNPB)-OH (**8**) as a *threo/erethro* mixture with a total yield of 19%. It is important to note, that loss of more than 50% in yield in the final exchange of protecting groups could have been avoided by use of commercially available Fmoc-L-Glu-OᵗBu in the STEGLICH esterification. This has been considered in following synthesis strategies.

3.1.2 7-Diethylamino-4-methylcoumarin (DEACM) protected amino acids

After NVOC and DMNPB protected SNARE-analogs were tested in fusion experiments and were found to be released too slowly (see section 3.3.3), a different type of caging group had to be chosen. DEACM was predicted be a good choice, especially to substitute NVOC, for its high quantum yield, high excitation wavelength (DEACM-cAMP, Φ = 0.24, 395 nm) and ultrafast release.[186,187] Uncaging can be efficient at wavelengths as high as 420 nm making it suitable for future microscopic applications as many confocal microscopes are equipped with a 405 nm laser. Furthermore, DEACM had already been established for caging both amines and carboxylic acids.[188,189]

In the mechanism of photolysis which is accepted to be common for (coumarin-4-yl)methyl esters (Scheme 3.5)[190,187], water plays a critical role. The heterolytically cleaved methylene-O-bond (l) will recombine if not scavenged by a nucleophile, for example water.

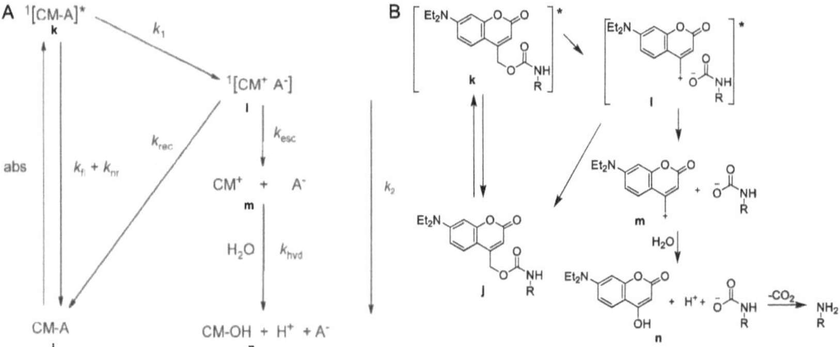

Scheme 3.5: A) Proposed mechanism of photocleavage for (coumarin-4-yl)methyl esters. In photoexcited k heterolytic bond cleavage can occur (k₁) or deactivation by fluorescence or nonradiave procecces (kᵣ + kₙᵣ). Recombination of l (kᵣₑc) competes with solvent separation (kₑₛc) and subsequent reaction with water (kₕᵥd). Published in: J. Phys. Chem. A 2007, 111, 5768-5774. Copyright © 2007 American Chemical Society.[190] B) Translation of the proposed mechanism to DEACM caged amine. Final decarboxylation reveals the free amine.

As the utilization of DEACM caged glutamic acid in SPPS had previously been reported[188] and DEACM caged lysine was not yet literature known, first, Fmoc-L-Lys(DEACM)-OH (**14**) was synthesized (Scheme 3.6) and tested with regard to SPPS and photocleavage. Following a synthesis procedure by ZHANG *et al*,[191] 7-Diethylamino-4-methylcoumarin (**15**) was oxidized with SeO_2 in xylene by heating to reflux for two days and the intermediate aldehyde was reduced to alcohol **16** with sodium borohydride. The use of these specific experimental conditions is emphasized, as other procedures with lower boiling point solvents and longer reaction times have been published,[192,193] but this procedure gave the best effort to yield ratio. Also, upscaling to more than 5 g is not recommended as selenium side products accumulate at the glass walls and stirring rod, interfering with stirring, and yields are reduced. Activation of the alcohol and conjugation to Fmoc-L-Lys-OH was achieved analogous to compound **4**.

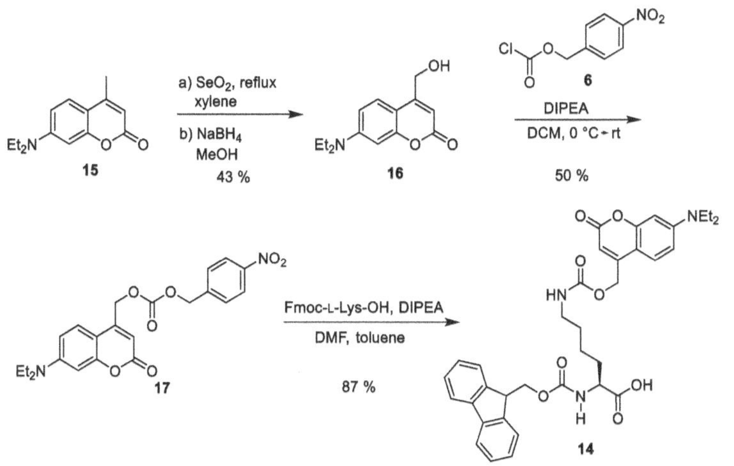

Scheme 3.6: Synthesis route to produce Fmoc-L-Lys(DEACM)-OH.

Short test peptide **18** (Scheme 3.7) was synthesized by manual SPPS and no major side reactions were observed. The caged peptide was used to determine photocleavage efficiency under the available experimental setup (method a, section 7.1.5). Owing to the high extinction coefficient of DEACM (16000 mol^{-1} cm^{-1})[194] a lower concentration was traceable by UPLC. Furthermore, considering the photorelease mechanism, the uncaging was expected to proceed best in aqueous medium so a 3.6 µM solution in HEPES buffer pH 7.4 was tested (see the appendix, Figure-A 3). Within 1 min, the caged peptide was consumed and the formation of photocleavage byproduct DEACM-OH was completed.

Additionally, peptides **19-21** (Scheme 3.7) were synthesized, to determine whether the cage would be available for photocleavage if the coiled coil was formed. The coiled coils were also chosen to test cleavage with a hand-held 405 nm 100 W laser pointer (method c, section 7.1.5), that would allow to condense the cleavage procedure in the fusion experiments. Uncaging proved to be complete within one minute (see the appendix Figure-A 4).

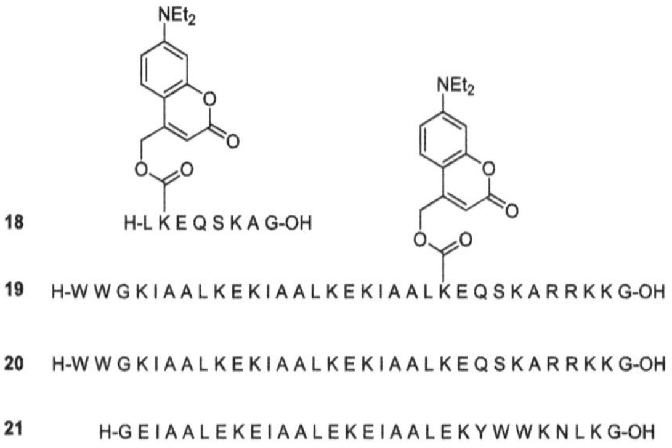

18   H-L K E Q S K A G-OH

19   H-W W G K I A A L K E K I A A L K E K I A A L K E Q S K A R R K K G-OH

20   H-W W G K I A A L K E K I A A L K E K I A A L K E Q S K A R R K K G-OH

21   H-G E I A A L E K E I A A L E K E I A A L E K Y W W K N L K G-OH

Scheme 3.7: Overview of test peptides used to assess photocleavage efficiency of DEACM caged lysine.

DEACM caged glutamic acid was synthesized by STEGLICH esterification of DEACM-OH (**16**) and Fmoc-L-Glu-O^{t}Bu as depicted in scheme 5.8. Acidic deprotection of the *C*-terminus produced Fmoc-L-Glu(DEACM)-OH (**23**) in an overall yield of 50%.

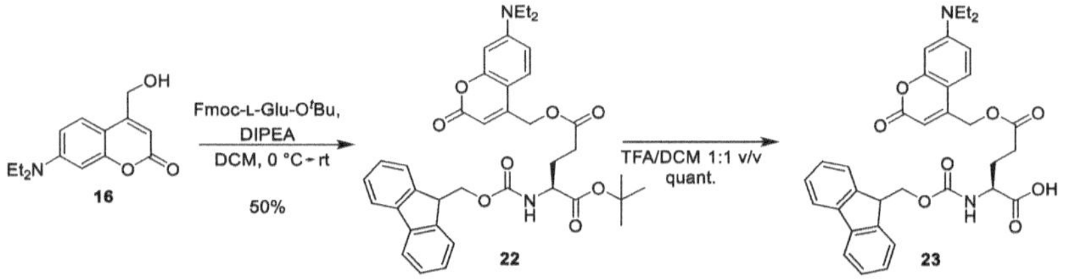

Scheme 3.8: Synthesis of Fmoc-L-Glu(DEACM)-OH.

3.2 Synthesized SNARE-mimicking peptides

Following the considerations illustrated in the beginning of this chapter, a selection of caged K3Sx and E3Syb derivatives were synthesized by microwave assisted SPPS. Target peptides are summarized in Table 3.1. K3Sx (**25**) and E3Syb (**2**) were needed for positive control experiments.

Table 3.1: Overview of peptides intended to assess the modulating effect of common PPGs in fusion experiments. Caging groups are indicated by bold lettering and placed on the right of the amino acid they are attached to. Peptides marked with green checkmarks were successfully synthesized and purified by HPLC using the conditions described in chapter 4. Synthesis of peptide 30 failed.

Molecule entry	Name	Sequence	
25	K3Sx	H-WWG(KIAALKE)₃-QSKARRKKIMIIICCVILGIIIASTIGGIFG-OH	✓
2	E3Syb	H-G(EIAALEK)₃-RKYWWKNLKMMIILGVICAIILIIIIVYFST-OH	✓

26	K3(NVOC)Sx	H-WWG(KIAALKE)$_2$KIAALK(**NVOC**)E-QSKARRKKIMIIICCVILGIIIASTIGGIFG-OH	✓
27	K3(NVOC)$_2$Sx	H-WWG(KIAALKE)$_2$K(**NVOC**)IAALK(**NVOC**)E-QSKARRKKIMIIICCVILGIIIASTIGGIFG-OH	✓
28	E3(DMNPB)$_2$Syb	H-G(EIAALEK)$_2$E(**DMNPB**)IAALE(**DMNPB**)K-RKYWWKNLKMMIILGVICAIILIIIIVYFST-OH	✓
29	K3(DEACM)$_2$Sx	H-WWG(KIAALKE)$_2$K(**DEACM**)IAALK(**DEACM**)E-QSKARRKKIMIIICCVILGIIIASTIGGIFG-OH	✓
30	E3(DEACM)$_2$Syb	H-G(EIAALEK)$_2$E(**DEACM**)IAALE(**DEACM**)K-RKYWWKNLKMMIILGVICAIILIIIIVYFST-OH	X

Synthetic principles described in chapter 2 could be successfully applied to the synthesis of the peptides listed in Table 3.1, except for E3(DEACM)$_2$Syb (**30**). Here, synthesis failed with major truncation products accumulated around the coupling positions of caged glutamic acid. At the time of synthesis, it was assumed that incomplete coupling caused by a combination of peptide aggregation on resin and steric hindrance of the caged building block sealed the fate of the peptide. However, considering the observations made in section 4.2, pyroglutamate formation in the successive coupling step is more likely.

3.3 Assessment of fusogenicity

3.3.1 Total lipid mixing assay

E3Syb and K3Sx derivatives caged by literature known PPGs (Table 3.1) were tested for their fusogenicity by a bulk total lipid mixing assay which exploits the physical phenomenon of fluorescence resonance energy transfer (FRET).[121] When the emission spectrum of a donor fluorophore overlaps with the absorption spectrum of an acceptor fluorophore a nonradiative transfer of energy is possible.[195] The efficiency of FRET is inversely proportional to the sixth power of the distance between the fluorophore pair making it highly sensitive to small distance changes:

$$\eta_{FRET} = \frac{1}{1 + (r/R_0)^6}$$

where η_{FRET} is the FRET efficiency, r is the distance between the fluorophores and R_0 is the FÖRSTER radius of the FRET pair which is defined as the distance at which η_{FRET} is 50%.[195]

Donor dye (NBD) and acceptor dye (lissamine rhodamine (Rh)), attached to the headgroups of phospholipids, are located in one population of liposomes at a concentration appropriate to produce efficient FRET (Figure 3.2). Upon merger of the liposomes with an unlabeled population, their lipids mix and the average distance between the dyes increases. This distance change can be read out by the increase of donor emission or the decrease of acceptor emission. Increase of donor emission is a more sensitive and accurate representation of the distance change, as in donor emission direct excitation can contribute to the total fluorescence value. Without modification, the assay cannot distinguish between full fusion of the liposomes and hemifusion where only the outer leaflet of lipids is mixed. However, deactivation of NBD by reduction with dithionite renders the merger of outer leaflets invisible and thus inner lipid mixing (ILM) can be assessed.[196]

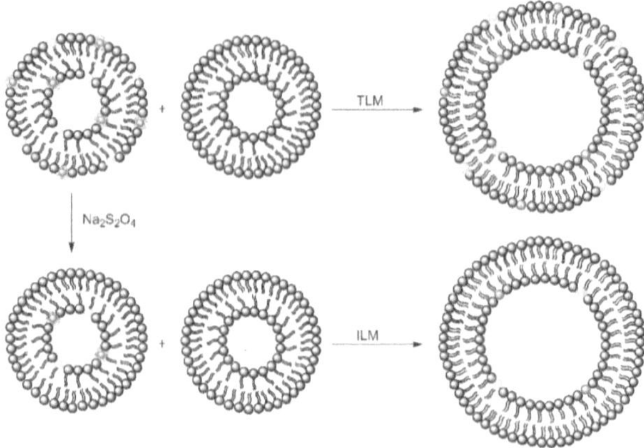

Figure 3.2: Simplified representation of bulk lipid mixing assays based on FRET.

For the peptides synthesized in this work, only total lipid mixing (TLM) experiments were carried out. Experimental parameters and procedures were employed in accordance with experiments from previous works.[134,197] Large unilamellar vesicles (LUVs) with the lipid composition DOPC/DOPE/cholesterol (50/25/25) were prepared following literature protocols[198] by extrusion of rehydrated and homogenized peptide/lipid films that were obtained by direct mixing of peptide and lipid stock solutions and drying in an N_2 stream. Labeled vesicles were doped with NBD-DOPE (1.5 mol%) and Rh-DOPE (1.5 mol%) while preserving the total DOPE concentration. E3Syb based peptides were reconstituted into the labeled population and the K3Sx based peptides were located in the unlabeled vesicles each with a peptide to lipid ratio of 1:200. Labeled and unlabeled vesicles were mixed at a ratio of 1 to 4 to obtain a final lipid concentration in the cuvette of 180 µM (calculated from the theoretical lipid content of homogenized vesicle emulsions before extrusion). Vesicle preparation and fusion experiments were carried out in HEPES buffer (20 mM HEPES, 100 mM KCl, 1 mM EDTA, 1 mM DTT) at pH 7.4.

In a time course measurement, NBD was excited at 460 nm and change of fluorescence followed at 535 nm over 20 min for experiments without uncaging and up to 120 min for experiments with uncaging. With increasing number of protecting groups the slope of the fusion curves was expected to become less steep or at best to follow the same course as the control experiment if the protecting groups were efficient at reducing coiled coil interactions. As fusion measurements are highly influenced by small differences in liposome preparation, for promising combinations, triplicates of the measurements were performed with vesicles obtained from extruding separate peptide/lipid films.

3.3.2   TLM of fusion pairs with one NV caged peptide

Total lipid mixing (TLM) experiments for peptides **2** + **25**, **2** + **26**, and **2** + **27**, were performed with **2** + **2** as negative control (Figure 3.3 A). The caged peptides showed no significant underperformance compared to the uncaged combination. **2** + **26** even exhibited a steeper increase of NBD fluorescence than **2** + **25**, which could be a peptide-related effect but is more probable to be caused by a higher insertion rate of the peptide or less loss of peptide/lipid film during proteoliposome preparation by extrusion. Despite limited evidence for the effective inhibition of coiled coil recognition in the peptide combination **2** + **27** (Figure 3.3 B), the peptide pair was investigated again regarding changes induced by UV irradiation (method a, see section 7.1.5). The fluorescence change was followed over a period of 2 h and the samples were irradiated at different time points for 10 s throughout the measurement. The irradiation time was kept short to be able to measure quickly after the trigger was set and to limit photobleaching of the fluorophores, even if the photorelease was probably incomplete. To account for photobleaching in the fusion curves the difference between last datapoint before and first datapoint after irradiation was added to all values after irradiation. The fusion curves after irradiation are corrected for photobleaching. In **2** + **27** (1) the phototrigger was applied after 1200 s for 10 s. A slight increase in slope of the fusion curve could be observed after irradiation. In the remaining experiments the curve progression was allowed to converge towards an equilibrium to observe a more distinct change of slope. In **2** + **27** (2) the irradiation appeared to cause an increase in fusion efficiency, while **2** + **27** (3) showed no change in curve progression after irradiation.

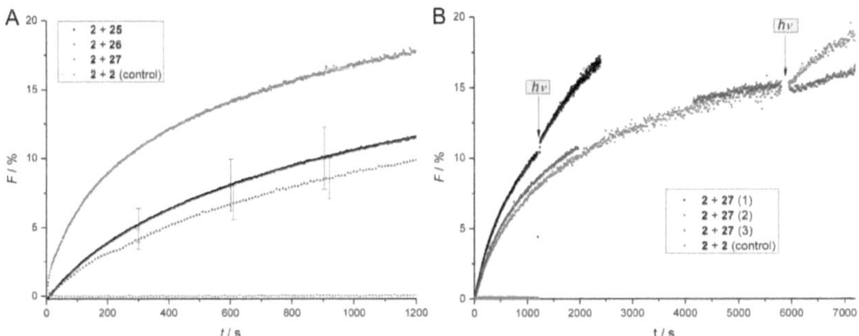

Figure 3.3: A: TLM of 2 + 25, 2 + 26 and 2 + 27 with 2 / 2 as negative control. Fusion curves 2 + 25 and 2 + 27 are an average of three repetitions. 2 + 26 was measured once. B: TLM of 2 + 27, three repetitions. The samples were irradiated for 10 s by method a) at different timepoints. The light blue line accounts for a measuring error of the spectrofluorometer during the measurement of the blue curve.

Overall, the combination of **2** + **27** was assessed to have a minimal modulating effect on the fusion efficiency that could be increased by photocleavage of the NVOC PPGs. However, the aim to suppress fusion activity with the PPGs was not achieved with this combination of peptides, so in subsequent experiments a more aggressive approach was considered.

3.3.3   TLM of fusion pairs with two peptides caged with NV based PPGs

The effect of removing two positive charges in K3(NVOC)$_2$Sx (**27**) did not produce the desired magnitude of reduction in fusion efficiency. It was then considered to modify the complementary heptad in E3Syb by introducing PPGs at the glutamic acids in e and g positions. The alteration would remove two more charges but more importantly introduce more bulk between the coiled coils as the glutamate residues are shorter than lysine residues and therefore less flexible. Thus, sterically demanding PPGs in these positions have the potential to insert between the coiled coils and disturb hydrophobic core formation. For this section, all experimental work has been executed by HOA NAM NGUYEN as part of his bachelor book.[185]

As control experiment, this time a combination of E3(DMNPB)$_2$Syb (**28**) in labeled vesicles with empty vesicles were measured. This was expected to represent the vesicles not interacting rather than being electrostatically repulsed by each other through the negatively charged peptides. Instead of the NBD fluorescence staying constant after the addition of unlabeled liposomes and throughout the measurement a significant increase in donor fluorescence was observed in the first 120 s of the measurement (Figure 3.4). This effect has also been noticed by HUBRICH and WEHLAND from the DIEDERICHSEN group who used analogous control experiments.[170,147] It could be attributed to the high sensitivity of NBD fluorescence to polarity changes which is given here by the introduction of a fourfold excess of lipids. The vesicles in this experiment can come in close proximity by BROWNian motion in contrast to the previous control where the proteoliposomes with the negatively charged peptide reconstituted in both populations would repulse each other. This initial gain in fluorescence was observed to be variable and is possibly a representation of absolute lipid concentration fluctuating by lipid loss in extrusion. The initial jump was followed by a slow, approximately linear increase in fluorescence. To test the response of the non-fusogenic proteoliposome/liposome mixture to light, irradiation for 120 s was performed. The length of UV exposure was expanded for this set of experiments to be able to completely release all PPGs. Upon continuation of the measurement a small jump in fluorescence was noted, possibly an artefact from opening and closing the measurement chamber, followed by linear, almost flat progression of the curve at about 3.5% fluorescence.

The combination of **28** + **27** was then tested in the TLM assay (Figure 3.4), mostly conserving the parameters of previous measurements. The fusion curves quickly reached a plateau similar to the curve shape of the control. In all three repeats, after about 120 s only a very shallow increase in F was monitored, indicating that the vesicles did not merge. After 1300 s the samples were irradiated for 120 s (method a), see section 7.1.5) to ensure complete release. Continued monitoring showed a significant increase of on average 3% in 1200 s after irradiation in donor fluorescence in all measurements, following the shape characteristic for fusion curves. When comparing this value of fluorescence increase to the pairing of **2** + **25** which had 10% F after 1200 s, the initial jump by polarity change must be taken into account. In Figure 3.3 the contribution of the fluorescence change by polarity shift cannot be quantified exactly as it is superimposed by fluorescence dequenching. Considering the control experiment, a generous contribution of up to 3% may be assumed. This would still leave a discrepancy of 3% between the unprotected **2** + **25** (Figure 3.3 A) and **28** + **27** after uncaging despite long irradiation times. On the other hand, the fluorescence values after irradiation are affected by photobleaching, the absolute value of which seems to be variable and was only corrected for by the minimal amount, to reach the fluorescence value before irradiation.

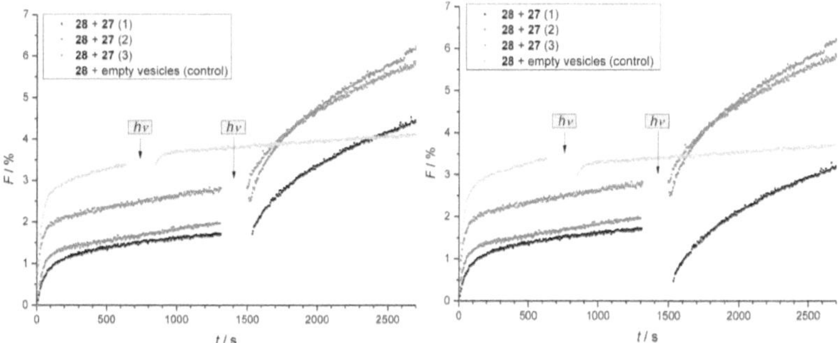

Figure 3.4: TLM of 28 + 27 and 28 + empty vesicles as negative control with (left) and without (right) correction for photobleaching.

In conclusion, the pairing two PPGs each in the membrane-proximal heptad of the recognition motifs produces a good inhibition of fusogenic peptides E3Syb and K3Sx. Photorelease by UV light is slow and does not completely restore function verified by fluorescence dequenching. Yet, the direct quantitative comparison to the original system is unreliable. Long deprotection times not only affect fluorescence by photobleaching of the dyes, but also obstruct access to the data that cannot be collected during irradiation. Therefore, the time required for uncaging was to be shortened by PPGs that would cleave more efficiently and a less time-consuming irradiation setup.

3.3.4 TLM of fusion pairs with one peptide caged with DEACM

A different type of PPG, that would be uncaged faster had to be found. DEACM was estimated to be a good choice for its faster photolysis and absorption maximum at ~380 nm. Synthesis of both **29** and **30** was attempted but could only be achieved for **29**. Although the effect of two removed charges in only one of the peptides was expected to be minimal considering the results of previous measurements, **29** was tested in TLM with **2** as fusion partner (Figure 3.5).

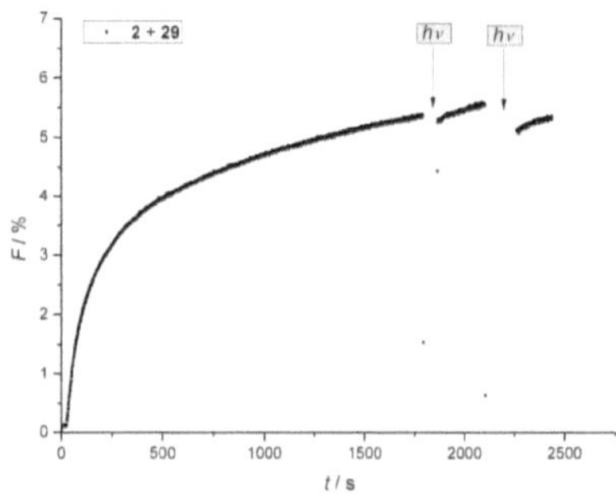

Figure 3.5: TLM of E3Syb/K3(DEACM)₂Sx, one measurement. The curve is not corrected for photobleaching.

The first 1700 s show a curve shape comparable to the positive control (E3Syb/K3Sx, Figure 5.2) with about 5% increase of NBD-fluorescence after 1200 s compared to E3Syb/K3Sx with 12%. The fusion efficiency seems to be hampered here, however, the difference lies within the conceivable fluctuations of the experiment. To assess the inhibitory effect of the caging groups on fusogenicity, the curve progression after irradiation is more meaningful. First, a 1 min of irradiation by method c) was performed. The hand-held laser pointer could be directly applied to the sample upon opening of the measurement chamber. As the time needed before to remove the cuvette from the spectrofluorometer and transport it to the Hg lamp and back could be saved, 60 s of irradiation was acceptable with regards to access to datapoints. From the assessment of photocleavage properties (see section 3.1.2), it can be concluded, that all DEACM would be released by this time. Positively, only very mild photobleaching was notable. However, the curve progression showed only a slight increase of slope in fluorescence. The following minutes of the measurement were therefore used to observe the effect of prolonged irradiation on NBD fluorescence. Additional 150 s of irradiation lowered F from 5.5% to 5.1% by photobleaching, but a change of slope could still not be detected.

3.4 Conclusion

In this section the fusogenicity of synthesized E3Syb and K3Sx derivatives with literature known PPGs was assessed. Photocleavable protection of four amino acids in the membrane-proximal heptads of the recognition units by NV based PPGs could inhibit lipid mixing in LUVs measured by total lipid mixing, but this proved to be unfavorable in time-resolved measurements due to the prolonged irradiation times required for deprotection. DEACM was envisioned as an alternative PPG with more suitable uncaging kinetics. Transfer of the previously identified caging pattern could however not be achieved as only K3(DEACM)₂Sx (**29**) could be synthesized by SPPS. The reasons for synthesis failure in E3(DEACM)₂Syb (**30**) were not determined but can likely be traced back to pyroglutamate formation after coupling of the building block Fmoc-L-Glu(DEACM)-OH. DEACM uncaging conditions in TLM of **2** + **29** were observed to be more benign with regards to photobleaching of the fluorophores required for TLM.

The number of PPGs that was needed to inhibit fusogenicity was presumably so high, because the electrostatic interactions that could be addressed by common PPGs are not primarily responsible for the strong attraction of E3/K3 coiled coils. We therefore aimed to develop a photoreversible protection strategy that would more precisely target the interface between the recognition units.

4 Development of a novel photocleavable protection strategy for coiled coil interactions

In chapter 3, the attempt to control coiled coil interactions by targeting charged amino acids in positions e and g was described (see Figure 4.1 D for a helical wheel representation). However, amino acids leucin and isoleucine in positions a and d form a tightly packed hydrophobic core at the contact area between the peptides which provides a major contribution to stabilizing the left-handed supercoil.[199] Removing electrostatic attractions from the equation can be straightforward by using widely established caging groups.[29] For leucine and isoleucine on the other hand, no simple protection strategy was found. It was envisioned that introduction of a bulky entity in between the coiled coils might disturb the formation of a hydrophobic core.

To introduce said bulk and secure it in place, stapled peptides were taken as an inspiration. Covalent connection of two (usually unnatural) amino acid residues is a technique used to stabilize secondary structure in alpha helix forming peptides[200,201] and has recently been applied to investigate drug targets for protein-protein interactions.[202,203] All-hydrocarbon stapled peptides, bridged by olefin metathesis, have been presented by the group of VERDINE.[30,31,204] Olefin bearing non-natural amino acids were used for ring closing metathesis (RCM) by Grubbs I catalyst to span over one (i,i+3 and i,i+4) or two (i,i+7) helix turns (Figure 4.1 A) of the investigated peptides.[30,31,204] Different chain lengths (x) and stereochemistry (R or S) of the used olefin bearing amino acid influenced RCM yield and conformational stability, with $R_{i,i+7}S(11)$ and $R_{i,i+7}S(12)$ showing highest conversion and strongest helix stabilization. For the design of stapled peptides, it is commonly recommended to avoid positioning the staple over a known interaction site.[31,205] The opposite principle has been employed in the design of photocleavably cross-linked peptide 31. Peptide E3 was chosen over K3 for modification. Due to the side chain length of the glutamic acid residues a lower flexibility of the moieties was expected, possibly restricting the number of product conformations. DEACM^{allyl} protected glutamic acids (e and g on the helical wheel, Figure 4.1 D) in a terminal heptad of E3 were used to form a connection ($S_{i,i+6}S(14)$) spanning over hydrophobic amino acids leucine and isoleucine. As a stapled peptide, this was anticipated to enhance α-helicity for the single peptide but destabilize coiled coil formation in combination with K3. For comparison, peptide 32 with a linkage via lysine residues ($S_{i,i+7}S(20)$) in f position of the central and terminal heptads was pursued. This linkage was expected to enhance single helix stability without disturbing coiled coil formation. In both peptides changing α-C stereocenters of the amino acids was out of question, as eventually, the native peptide had to be released.

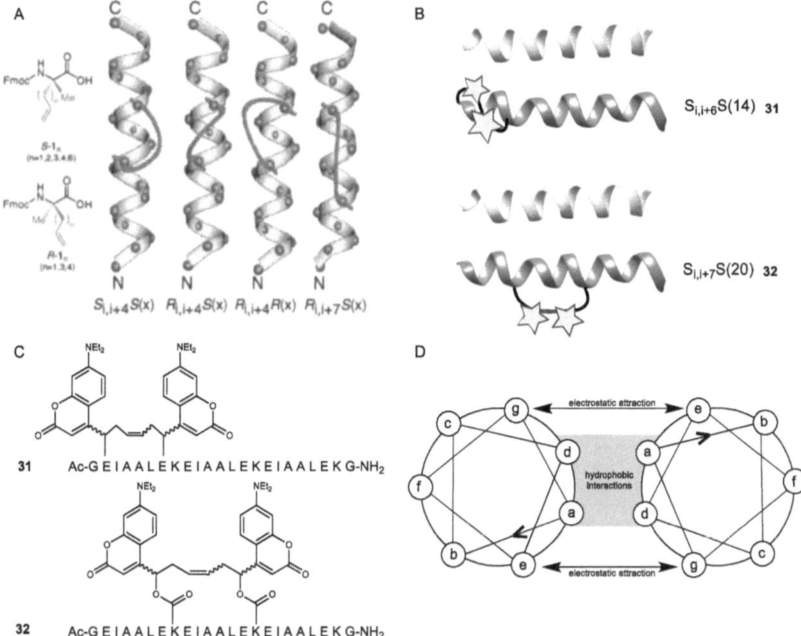

Figure 4.1: A: All-hydrocarbon stapled peptide nomenclature as presented by SCHAFMEISTER et al.[30]. B: Schematic illustration of E3/K3 coiled coil with E3 photocleavably cross-linked via glutamate residues (top) to weaken hydrophobic interactions and via lysine residues (bottom) as putative non-interfering comparison. C: Sequence and structure of photocleavable cross-link of 31 and 32. D: Helical wheel representation of geometry and interactions in coiled coils.

4.1 Building blocks for photocleavable cross-link

A new DEACM based photoactive group, that can be used simultaneously as PPG and PCL was developed for the application in reversible intramolecular cyclization. For this purpose, the DEACM base structure was equipped with an allyl linker that can be selectively addressed in olefin metathesis. GRUBBS I catalyst has been used for ring closing metathesis of olefins and alkynes in resin bound peptides before.[31,30,204,206] The so called stapled peptides have been used to provide conformational stability to peptide drugs and thereby enhance their performance.[206] Photo-switchable staples based on cis/trans photoisomerization have been published before to reversibly control peptide secondary structures.[207,208] To our knowledge, no photoreleasable stapling technique has been presented to this date that recovers the native sequence after irradiation.

To synthesize DEACM^allyl-OH (**33**, Scheme 4.1), DEACM (**15**) was oxidized with SeO_2 and the aldehyde (**34**) was isolated. The aldehyde was further reacted with allyl stannane **35**, a very toxic but very effective allylation reagent, to obtain **33** in 46% overall yield as a racemic mixture.

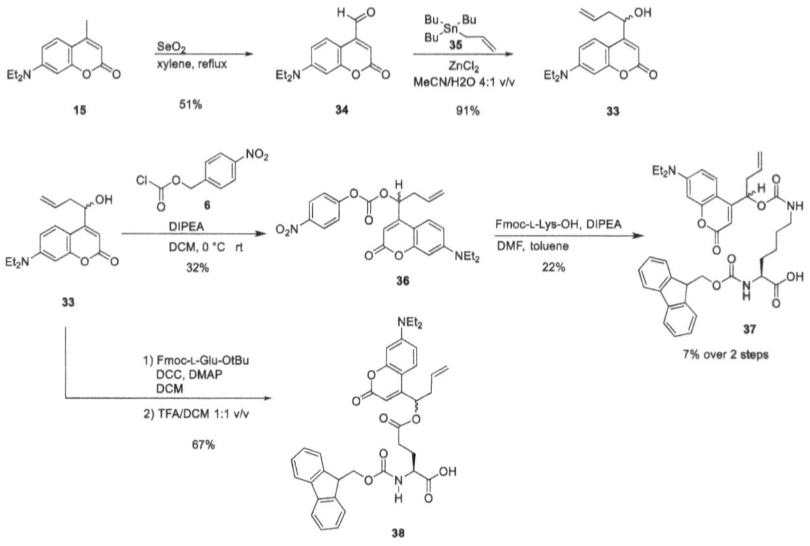

Scheme 4.1: Synthesis of amino acid building blocks 37 and 38 for phocleavable peptide cross-linking.

The alcohol was then used with the established conjugation procedures to obtain Fmoc-L-Lys(DEACM^{allyl})-OH (**37**) with 7% yield and Fmoc-L-Glu(DEACM^{allyl})-OH (**38**) with 67% yield.

PPG/PCL conjugated lysine **37** was further transformed to assess photochemical properties of the new DEACM derivative. Water soluble compounds **39** and **40** were synthesized to measure extinction coefficients of **39** and **40** and photocleavage rates and byproducts of **40**. **39** was produced from **37** by Fmoc deprotection in solution. Compound **40** was obtained from GRUBBS I catalyzed metathesis of **37** in solution and subsequent Fmoc deprotection.

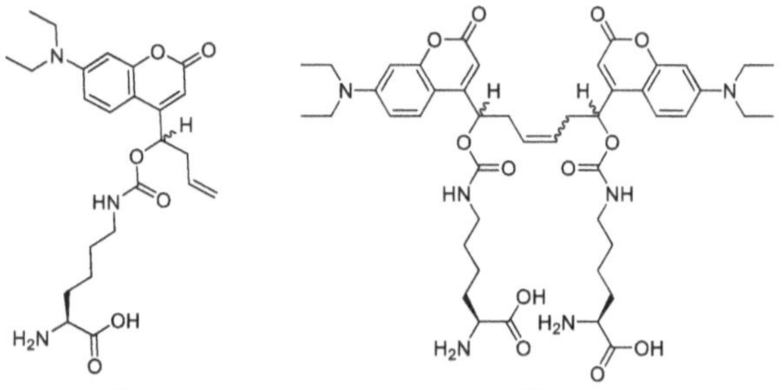

Figure 4.2: Water-soluble compounds 39 and 40 used to investigate photophysical and photochemical properties of the novel PCL.

Extinction coefficients ε_{max} for **39** and **40** were obtained from UV measurements of dilution series in MeCN and in water (see the appendix, Figure-A 8 and Figure-A 9). The absorption maximum was determined at 380 nm for both compounds, thus, ε_{380} was measured. For both compounds a mixture of all obtained stereoisomers was used. ε was obtained from the slope of absorption plotted against concentration.

Table 4.1: Extinction coefficients measured from dilution series of compounds 39 and 40 in MeCN and H₂O.

Compound	ε_{380} [M^{-1} cm^{-1}] in MeCN	ε_{380} [M-1 cm-1] in H_2O
39	17018 ± 188	16181 ± 348
40	25972 ± 377	23121 ± 395

While the ε_{380} of **39** was similar to ε_{385} of DEACM,[194] ε_{380} of metathesis product **40** could not have been predicted by ε summation of the two chromophores. This reduction in extinction coefficient suggests $\pi\pi$-stacking of the chromophores enabled by the proximity within the molecule and probably varies between the different stereoisomers. Nonetheless, the average values obtained from measuring the mixtures of isomers were used in all applications.

40 was used as a model to examine uncaging chemistry and rates of (DEACM)$_2$^{butenyl} bridged compounds. For DEACM the uncaging byproduct is known to be well defined (Scheme 3.5),[187] so a similar behavior was anticipated. Uncaging products were identified by LC-MS analysis. A 100 µM solution in ultrapure water was uncaged with irradiation setup c) (see section 7.1.5) for 30 min and samples drawn at different time points were analyzed by LC-MS (corresponding LC-MS analysis is shown in the appendix Figure-A 5, Figure-A 6, Figure-A 7). Over the course of the irradiation the formation of precipitate was noticed, presumably composed of insoluble uncaging byproducts. Peaks at different retention times with the same mass could be attributed to the presence of multiple stereoisomers. The assignment is summarized in Table 4.2.

Table 4.2: Molecular ion peaks calculated from LC-MS signals obtained by uncaging of 40.

	0 min			3 min			30 min	
R_t [min]	m/z	Δ m/z	R_t [min]	m/z	Δ m/z	R_t [min]	m/z	Δ m/z
			0.7	146.0	774.4	0.7	146.0	774.4
5.9	890.4	0	5.9	890.4	0	6.5	718.3	172
			6.5	718.4	172	6.8	672.4	218
			7.3	700.3	190.1	7.5	700.3	190.1
			7.5	700.3	190.1	7.9	546.3	344.1
			7.9	546.3	344.1	8.0	546.3	344.1
			8.0	546.3	344.1	8.2	528.4	362
			8.2	528.4	362	8.4	528.4	362
			8.4	528.4	362	8.7	500.3	390.1
			8.7	500.3	390.1	9.1	528.4	362
			9.1	528.4	362	9.5	528.4	362
			9.5	528.4	362			

At 0 min irradiation the peak at R_t = 5.9 min with m/z = 890.4 was detected which matches the educt. After 3 min of irradiation, the educt was still present and after 30 min it was completely consumed. Beyond that the compounds detected after 3 and 30 min mostly have coinciding m/z values and shall only be discussed qualitatively. In both analyses, the mass of free lysine, 146.0, could be identified in the elution peak at 0.7 min. The peaks with m/z = 718.4 and 546.3 have a $\Delta\,m/z$ of 172 and 344.1 from the educt respectively. The mass changes are in accordance with the photolysis mechanism of DEACM by heterolytic cleavage of one or both carbamate lysins and successive nucleophilic attack of water at the resulting carbocation. The link between coumarins stays intact and in place of the carbamate they bear a hydroxy function. Two possibly related peaks are m/z = 700.3 ($\Delta\,m/z$ = 190) and 528.4 ($\Delta\,m/z$ = 362). The molecules each have a $\Delta\,m/z$ = 18 from the previously discussed photolysis products which matches a water molecule. This could be a result of direct elimination of the carbamate unit accompanied by H^+ abstraction or first formation of the hydroxylated molecules and subsequent water elimination, in both cases creating an unsaturated bond conjugated to the neighboring coumarin system. Notably, there is no evidence that the same reaction takes place at the second hydroxy group. However, this could be explained by the reaction being disadvantaged due to insolubility of the product in aqueous environment. The last prominent pair of products has a m/z = 672.4 (Δ = 218) and 500.3 (Δ = 390.1). Starting from 700.3 and 528.4 a net loss of CO can be calculated. Plausible molecular structures for the photorelease of the DEACM-based crosslink are summarized in Scheme 4.2.

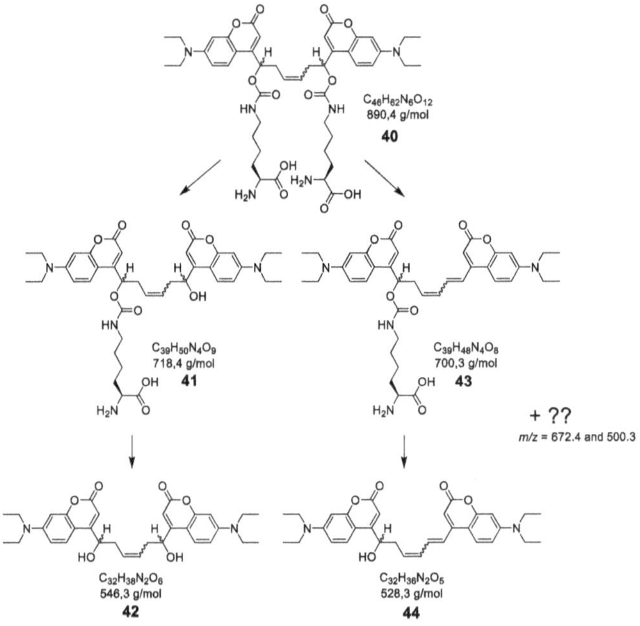

Scheme 4.2: Plausible molecular structures for the photolysis products observed in LC-MS.

To estimate the effect of concentration on the required irradiation times, corresponding uncaging experiments were performed at different concentrations (Figure 4.3). Considering the precipitate formation in the preceding measurement the evaluation of photolysis kinetics was performed with 0.1% pluronic F127 in H$_2$O to help solubilize the photolysis products and minimize scattering of the laser beam. For analysis, two types of events were considered. Disappearance of all lysin-carrying molecules, including intermediates **41** and **43** was representative for the full recovery rate of native peptide. In addition, release of at least one lysin and thus, reduction of peak area of only **40** reflects the ring opening rate in a cyclized peptide. At all concentrations except 115 µM, full deprotection was completed after 3 min and release of at least one lysine was accomplished after 2 min. Between 71 µM and 115 µM there could be a solubility threshold in the used medium, slowing down the photochemical reaction.

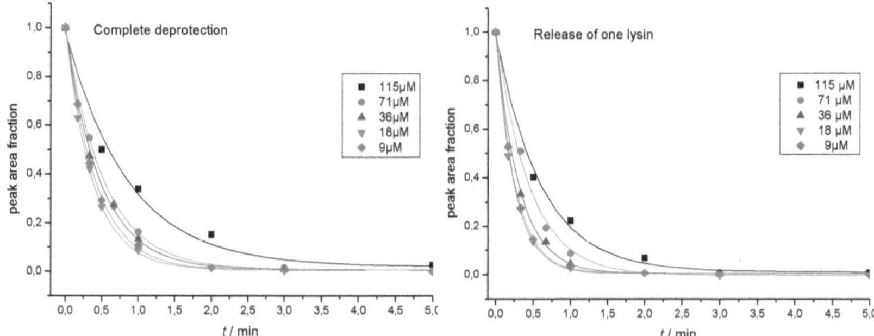

*Figure 4.3: Uncaging of **40** at different concentrations in 0.1% pluronic in water with uncaging setup c). Left: For complete deprotection the peak area of the educt and all identified intermediates still bound to one lysine were considered. Right: Only the peak area of **40** was plotted, representative of ring opening rate in a peptide.*

4.2 Synthesis of photocleavably cross-linked testpeptides

Synthesis of both pre-macrocyclization peptides **45** and **46** was attempted by automated SPPS using DIC/Oxyma as activators with exception of photocleavable building blocks **37** and **38** which were coupled manually with HATU/HOAt (Scheme 4.3). While **46** was obtained in good quality, very little product was obtained for **45** although the manual coupling steps had been followed by KAISER test and suggested complete coupling.

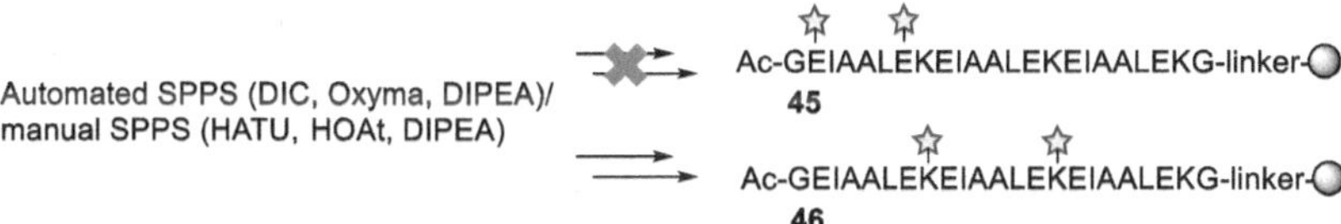

*Scheme 4.3: **45** could not be obtained from automated SPPS/manual SPPS. Synthesis of **46** was straightforward. The yellow stars represent DEACM^{allyl}.*

Therefore, a more thorough investigation of the reaction was undertaken. All steps before coupling the first caged glutamate (see Figure 4.4 (A)) were performed by automated

SPPS. All other steps were carried out manually and resin samples for test cleavage were drawn at different stages of the synthesis. After test-cleavage, the peptides were analyzed by UPLC and ESI-MS (Figure 4.4). The presence of several product peaks at later stages is owed to diastereomeric building block **38**.

From the progression of the synthesis it can be observed that neither the coupling of the caged glutamate (B) nor the Fmoc deprotection (C) produce any side product in significant amounts. Also, it appears that peptide cleavage at strongly acidic conditions does not contribute to the side product formation as the conditions were the same for every stage of the synthesis and side products were only verified at the later stages. Interestingly, at stage (C) only one product peak was observed despite expecting two diastereomers. Possibly the two products cannot be separated at the applied conditions, as the following test cleavage analysis shows the expected number of peaks again. The next test cleavage was performed only after four more full coupling cycles (D) so the conclusion may not be undisputable, yet, the detection of the 1822.0 g/mol peak at this stage suggested side product formation during the coupling of the first aa after (C), more precisely leucin. The mass corresponds to pyroglutamate formation under the loss of DEACM^{allyl}-OH. Detection of a single side product peak is plausible as will be discussed further down below. Coupling of the second caged glutamate (D) did not produce any further side product and the expected number of product peaks were detected. After finalization of the sequence and N-terminal acylation a side product with 2588.4 g/mol matching the mass of pyroglutamate formation from the second caged glutamate was measured. Furthermore, two additional peaks with molar weights 1864.0 g/mol and 2630.4 g/mol were detected and could be attributed to acylated pyroglutamates at both positions.

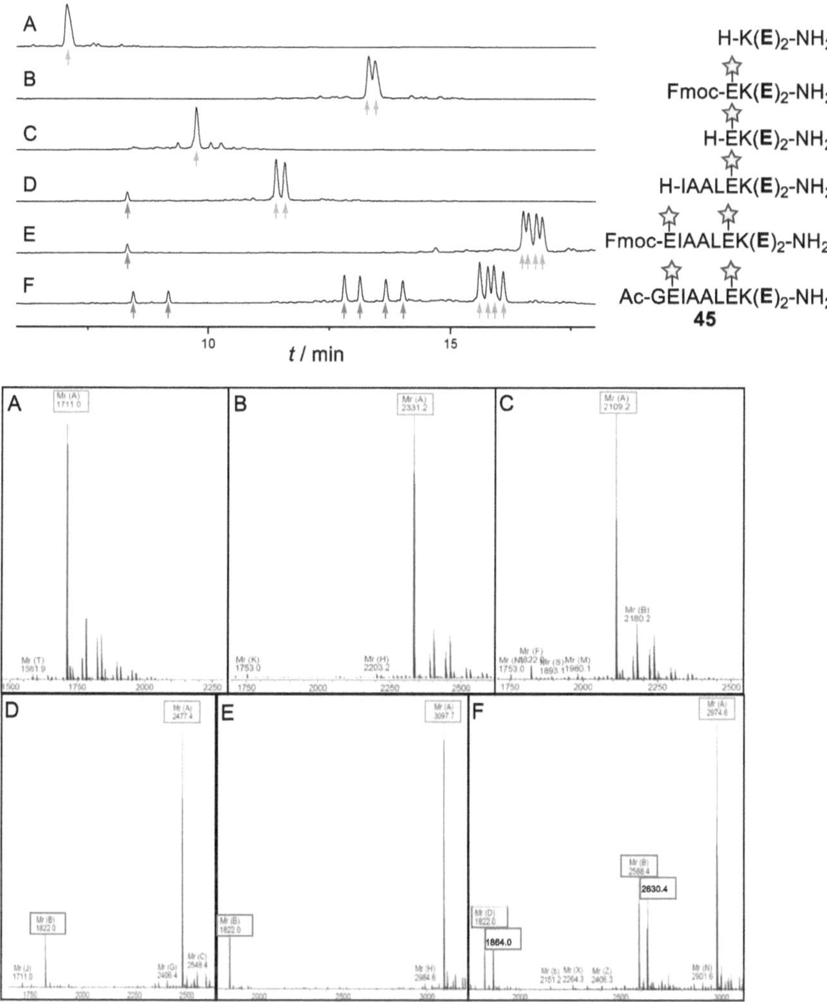

*Figure 4.4: Synthesis tracking of **45** with UPLC analysis (top) with solvent system II and a linear gradient of 10-80% B in 20 min and corresponding deconvoluted ESI MS spectra (bottom). Samples were drawn after completion of the indicated reaction stage and cleaved following SOP (7). (**E**)₂ signifies the two repetitions of EIAALEK and the yellow stars represent DEACM^{allyl}. Signals of target products for each stage are highlighted in green and identified side products are highlighted in red.*

Pyroglutamate formation in fully deprotected peptides in aqueous solution is known to occur in N-terminal glutamic acids and glutamines in mildly acidic conditions.[209] The reaction takes place by nucleophilic attack of the primary amine at the γ-CO and is concluded by release of H_2O or NH_3, therefore being promoted by presence of a mild acid providing the proton. Strong acids would completely protonate the amine leaving the electron pair unavailable for nucleophilic attack. In SPPS following Fmoc/'Bu strategy unwanted pyroglutamate formation from glutamate is usually not a problem. Typically, the side chain is protected with 'Bu, which can shield the carbonyl from nucleophilic attack by steric hinderance. With other protecting groups however, pyroglutamate formation has been known to occur, in those cases it is attributed to the Fmoc deprotection stage.[210]

In case of Fmoc-L-Glu(DEACM^{allyl})-OH (**38**) coupling, pyroglutamate formation in Fmoc deprotection could only be observed if 0.1 eq of formic acid or HOBt, both weak acids, were added to the 20 % piperidine in DMF solution as is common to suppress aspartimide formation (not shown). In the successive coupling step both the acid function of the amino acid and the activator additives HOAt in manual coupling or Oxyma in automated SPPS have the potential to catalyze the side reaction. It appears that in the competition between coupling and cyclization the coupling is favored if the coupling cocktail is allowed to incubate for a few minutes thereby preactivating the aa prior to addition to the resin. In automated SPPS, where the solutions of amino acid, DIC and Oxyma are combined in the reaction vessel to concurrently meet the free amines of on-resin peptides no product formation was observed. On the other hand, product formation was observed employing manual coupling with preactivation. Pyroglutamate formation terminates the peptide sequence and no further oligomerization can take place (Scheme 4.4). However, the amide-N is susceptible to acylation as can be seen from Figure 4.4 (F). The cyclisation products lose their stereo ambiguity and the UPLC peak number is halved as the undefined stereocenter is located at the cleaved DEACM^{allyl}-OH (**33**).

For future applications preactivation of the amino acid coupled after the modified glutamate should be the minimal precaution and the yield might be improved by use of more efficient activators. If full conversion at this step is crucial for a project preparation of an NHS-ester or a similarly activated amino acid derivative might be considered in place of *in situ* activation as to avoid the presence of any acid.

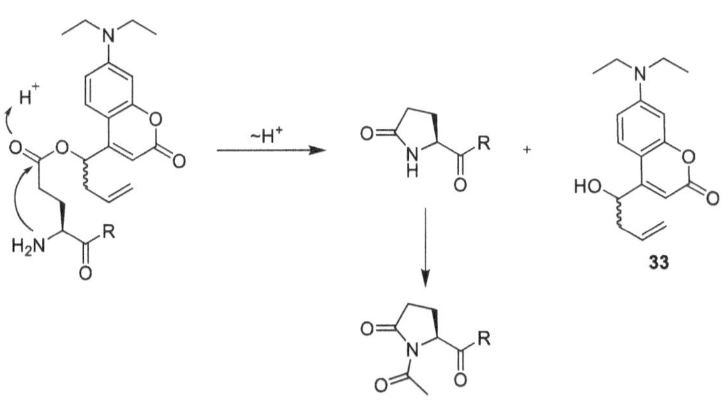

Scheme 4.4: Formation of cyclic pyroglutamate by nucleophilic attack and release of DEACM^{allyl}-OH catalyzed by acid. Pyroglutamate terminates the oligomerization - however, it can be acylated with Ac₂O.

Despite the synthetic obstacles in the linear peptide synthesis, both resin-bound target peptides **45** and **46** could be successfully synthesized. RCM was then performed taking advantage of Grubbs I catalyst. For **31**, common HPLC solvent systems III and IV (with 0.1% TFA) could not separate the product peaks from pyroglutamate peaks even at very flat gradients. In pre-purification with neutral buffer TEAA and acetonitrile the truncated peptides eluted distinctly after the product peaks. In a second semipreparative RP-HPLC step with system III a mix of diastereomeric product peaks was isolated with satisfactory purity (Figure 4.5 left). For **32**, standard purification methods could be applied to isolate two main product peaks (Figure 4.5 right).

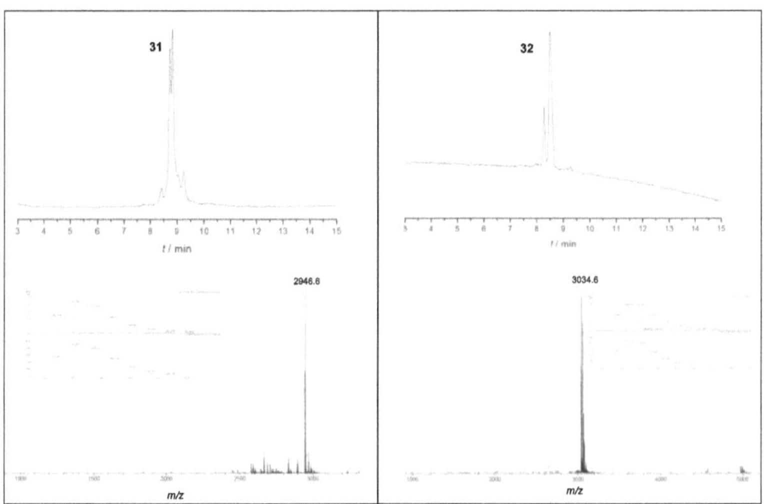

Figure 4.5: UPLC analysis and deconvoluted ESI-MS with HR-MS spectra of purified peptides 31 (left) and 32 (right). Chromatogramms were recorded with solvent system I and a gradient of 40 to 80% B in 15 min and a flow of 0.45 mL/min. Absorption was observed at 390 nm.

4.3 Properties of photocleavably cross-linked testpeptides

After isolation of **31** and **32**, some of their properties were investigated. First, it was tested if photolysis would release the native peptide without side reactions. For this, small amounts of peptide were dissolved in ultrapure water and irradiated by method b) (see section 7.1.5) for 2 min. The solutions were analyzed by UPLC before and after irradiation (Figure 4.6). For both photocleavably cross-linked peptides, release of the fully deprotected peptide Ac-G(EIAALEK)$_3$G-NH$_2$ could be verified my ESI-MS.

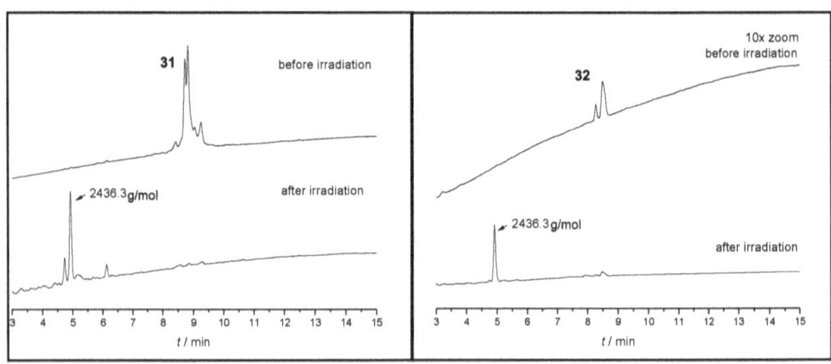

Figure 4.6: UPLC analysis of 31 and 32 before and after irradiation. Chromatogramms were recorded with solvent system 1 and a gradient of 40 to 80% B in 15 min and a flow of 0.45 mL/min. Absorption was observed at 215 nm. The newly formed peak was identified by mass spectrometry.

Circular dichroism (CD) spectroscopy is a common method for analyzing coiled coil interactions. Conformationally fixed chromophores of chiral molecules absorb right-hand and left-hand circular polarized light to different extents which can be measured as CD signal. Many oligomeric biomolecules and their structural mimetics produce characteristic CD spectra by which their secondary structure can be distinguished.[211–213] Coiled coil peptides E3 and K3 show the characteristic curve shape of α-helices, with equally intense minima at 208 nm and 222 nm and a maximum at 192 nm when combined at a 1:1 ratio.[214,215,212] However, when the individual peptides are isolated, they form random coils at neutral pH with a minimum at about 200 nm. Owing to this dichotomy, dissociation of the heterodimeric coiled coils at elevated temperatures can be read out by the gradual disappearance of the signal at 222 nm and the unfolding temperature T_{unfl} is equated with the melting temperature T_m of the dimer.

Cross-linked peptides **31** and **32** were analyzed by CD as solutions of the individual peptides and compared with an unlinked E3 derivative (Ac-GEIAALEW(EIAALEK)$_2$G-NH$_2$, **47**) (Figure 4.7 A). Furthermore, CD spectra of the E3 derivatives as 1:1 mixtures with a K3 derivative (Ac-GKIAALKW-(KIAALKE)$_2$G-NH$_2$, **48**) (Figure 4.7 B) were measured. The analysis was completed by unfolding curves of the complexes recorded by observing the CD signal at 222 nm between 5 and 95 °C. For all measurements the total peptide concentration was adjusted to 20 µM in 10 mM phosphate buffer pH 7.0, with the concentration being limited by the solubility of **32**. CD signals were normalized to mean residue ellipticity *[θ]*. **47** and **48** were kindly provided by PIRAJEEV SELVACHANDRAN[2].

Cross-linked peptides **31** and **32** were expected to show distinct signs of organization in comparison to unlinked peptides **47** and **48**. In fact, **31** does produce a signal of -12.1 deg cm^{-1} dmol^{-1} at 222 nm, that is significantly more intense than in the unlinked peptides (-3.9 deg cm^{-1} dmol^{-1} for **48** and -7.0 deg in **47**). Also, the minimum at shorter

---

[2] Institute of Organic and Biomolecular Chemistry, Georg-August-University of Göttingen

wavelengths is shifted to 205 nm. In **32** on the other hand, the only distinct difference to **47** was detectable by the shift of the minimum to 204 nm and a slightly less intense signal.

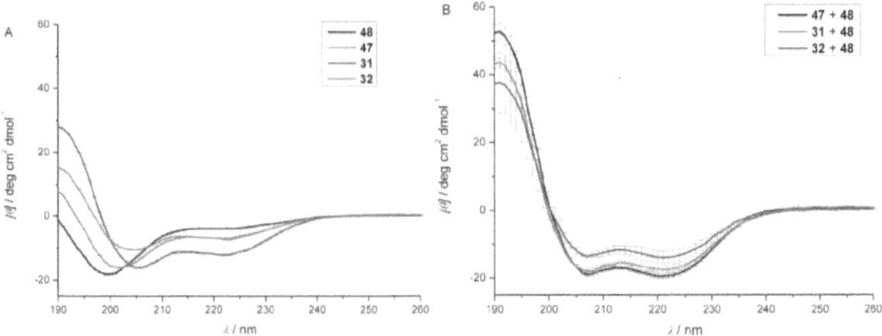

Figure 4.7: CD spectra of individual peptides 31, 32, 47 and 48 (A) and of 1:1 mixtures of E3 derivatives with 48 (B). Spectra of 20 μM solutions in 10 mM phospate buffer pH 7 were recorded at 25 °C.

In the 1:1 mixtures with **48**, **31** was anticipated to disturb dimer formation while the cross-link in **32** should not have influenced the coiled coil interaction. Differing from this prediction, CD-curves of **31 + 48** and **47 + 48** were nearly congruent, while **32 + 48** showed an overall less intense signal but with the same curve shape. This could indicate that in **31** with the cross-link introduced at the *N*-terminus, the former is flexible enough to move out of the hydrophobic core of the coiled coil and thus not interfere with the later. Alternatively, the coumarins might participate in the hydrophobic interactions rather than disturbing them. For **32**, the macrocycle produced by the cross-link consists of more units than the staples reported to have helix stabilizing effects.[30] This could produce unintended strain or a bend to the coiled coil of **32 + 48**, thus reducing helical content.

Considering the curve shapes of recorded CD spectra, unfolding behavior at elevated temperatures is more indicative of the stability of secondary structure introduced in the E3 derivatives by the cross-link than of the stability of coiled coil interactions (Figure 4.8). The calculated turning points are therefore indicated as T_{unfl} instead of T_m.

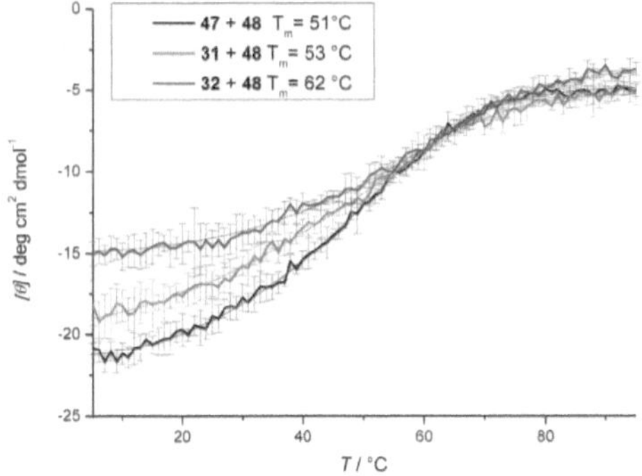

Figure 4.8: Thermal unfolding of 20 μM solutions in 10 mM phospate buffer pH 7 of 47 + 48, 31 + 48 and 32 + 48 followed via the CD-signal at 222 nm.

In conclusion, a photoreversible intramolecular cross-linking strategy has been developed, which can induce α-helical structure depending on the size of the formed macrocycle. To verify effects on coiled coil interactions, for future experiments either the central heptad should be targeted, or the peptide should be extended by a few amino acids to further reduce flexibility of the cross-link. Although significant effects of the photocleavable cross-link for soluble coiled coils could not yet be determined, a putative effect on fusogenicity of E3Syb/K3Sx is also not excluded. From a synthetic point of view, Fmoc-L-Lys(DEACM^{allyl})-OH (**37**) was much less problematic than the glutamic acid counterpart (**38**) and was therefore used to translate the new PPG/PCL strategy to K3Sx and synthesize a cross-linked K3Sx derivative.

4.4 Synthesis of a photocleavably cross-linked K3Sx derivative

Previous findings concerning synthesis and structure of photocleavably cross-linked peptides were translated to the synthesis of a SNARE analog derivative. The choice between targeting E3Syb and K3Sx was easily made by the argument of synthetic approachability. Fmoc-L-Lys(DEACM^{allyl})-OH (**37**) was found to be applicable to standard Fmoc-SPPS without major side reactions while Fmoc-L-Glu(DEACM^{allyl})-OH (**38**) was susceptible to pyroglutamate formation which was a major source of truncated sequences in **31** (see section 4.2). Thus, synthesis of K3((DEACM)$_2^{butenyl}$)Sx (**49**) was attempted, with the cross-link introduced via the lysine residues of the membrane-proximal heptad of K3.

Synthesis of the pre-macrocyclization peptide (**50**) was achieved analogous to previous peptides (chapter 2) by automated SPPS with the CarboMAX activation method and microwave assisted coupling at 90 °C, with the exception of Fmoc-L-Lys(DEACM^{allyl})-OH (**37**) (Scheme 4.5). Coupling of the photocleavable building block was performed manually (with HATU/HOAt as activators) to be able to recycle excess equivalents of the building block. For this synthesis, the *N*-terminus was supposed to be an amine, so no acylation was performed. The *N*-terminal Fmoc was left on the peptide to exclude side reactions of the amine in the following macrocyclization step. After successful formation of the intermediate **50** was confirmed by ESI-MS, ring-closing metathesis was performed on resin promoted by Grubbs I catalyst. After Fmoc deprotection and acidic cleavage crude **49** was obtained.

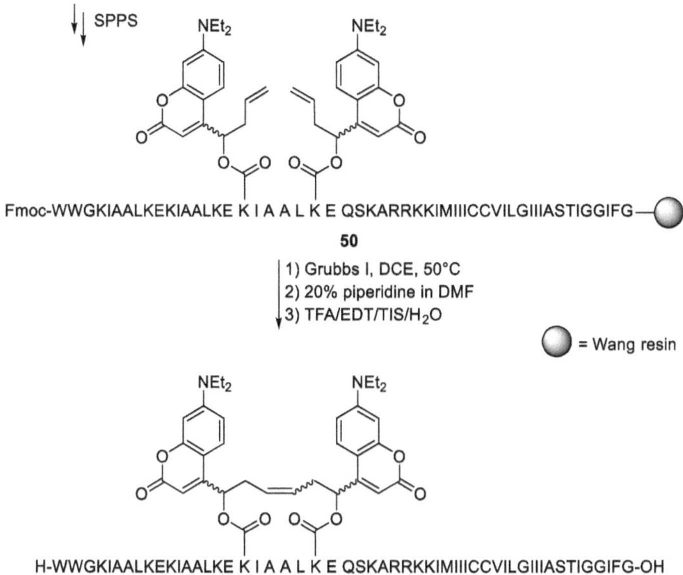

Scheme 4.5: Synthesis of photocleavably cross-linked K3((DEACM)$_2^{butenyl}$)Sx (49).

Because of the length and difficulty of the sequence, the crude product contained the target peptide only to a small percentage as could be determined by LC-MS (Figure 4.9). Employing the sample preparation methods described in chapter 2, purification of **49** could be achieved in a two-step RP-HPLC. Individual diastereomers were not resolved by chromatography, however, up to eight isomers are conceivable.

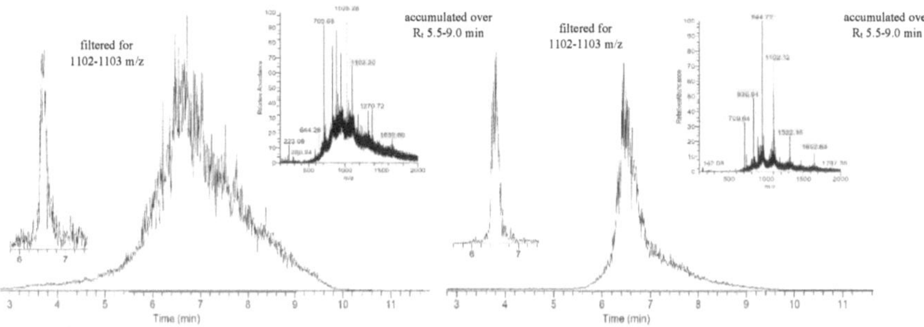

Figure 4.9: LC-MS analysis of K3((DEACM)$_2^{butenyl}$)Sx (49) crude (left) and after purification by HPLC. Mass spectra were accumulated over the time period marked with red and signals corresponding to the target peptide are highlighted in green.

4.5 Conclusion

In this section the development of a novel photocleavable intramolecular linker as a photocleavable protection strategy for coiled coils was described. The approach was inspired by stapled peptides developed in the group of VERDINE, which is why the developed linker is also referred to as staple. A PPG/PPL consisting of DEACM equipped with an allyl linker for olefin metathesis was designed and synthesized. The novel PPG was applied to the caging of the side chains of Fmoc-L-Glu-OH and Fmoc-L-Lys-OH. Extinction coefficients were measured and photocleavage behavior was investigated using soluble derivatives of the caged lysine monomer. For metathesis product **40** an unexpected reduction of extinction coefficient and a diverse side product ensemble after photocleavage was observed.

In the synthesis of peptides **31** and **32** the suitability of the amino acid derivatives for SPPS was assessed. While caged lysine could be incorporated without major side reactions, pyroglutamate formation could be identified as a major side reaction in the oligomerization step following the coupling of Fmoc-L-Glu(DEACM^{allyl})-OH (**38**). Despite synthetic challenges, after Grubbs I catalyzed ring-closing metathesis, two stapled derivatives of E3, representing one half of the recognition pair used in the SNARE analogs developed by MEYENBERG, could be isolated. One derivative was linked via two caged glutamic acids, spanning hydrophobic amino acids in the N-terminal heptad of the peptide and one peptide was linked via caged lysins on the site which is exposed to the solvent when a coiled coil is formed. Both peptides could be completely recovered to their uncaged form by irradiation with a 405 nm laser beam.

The effect of the staples placed in different positions of E3 on the secondary structure was investigated in peptides **31** and **32** by CD spectroscopy. In contrast to anticipated behavior, peptide **31** showed enhanced α-helicity compared to unstapled **47** when isolated but did not significantly interfere with coiled coil formation in combination with **48**. On the other hand, in **32** the effect of the staple on the isolated peptide was less distinct, but coiled coil formation with K3 was hampered. It was concluded that in **31** the positioning of the staple allows for high flexibility preventing an inhibiting effect on coiled coil formation while in **32** the staple may induce a bend of the peptide, thus reducing attractive interactions by a changed geometry.

The lessons learned from the synthesis of the peptides **31** and **32** were applied to the synthesis of photocleavably stapled K3Sx derivative **49**. Employing the HPLC conditions optimized for the purification of difficult peptides, **49** could be isolated with satisfactory purity. The peptide is now ready for future lipid mixing measurements analogous to section 3.3. If the peptide can be shown to efficiently inhibit lipid mixing and to be restored to its full function within a reasonable time frame, further fusion experiments can be performed to distinguish between mixing of the proximal lipid leaflets, mixing of proximal and distal leaflets, and content mixing. The peptide can then be used to study different stages of zippering as is observed in the native SNAREs. It is of interest how fusion intermediates are affected by exchanging amino acids in the linker region of the SNARE mimetics or varying lipid composition in the model membranes.

5 Tracking membrane buried and soluble antigens using artificial peptides

In the following sections the results of antigen binding studies are presented and discussed. Peptides required as antigen source were mostly provided by solid phase synthesis as part of the scope of this book. All *ex vivo* experiments were planned and carried out by FRANS BIANCHI, ELKE MUNTJEWERFF, MAXIM BARANOV AND SJORS MAASSEN in the VAN DEN BOGAART group[3]. As this book is presented with a focus on chemistry, detailed workflow concerning the handling of live cells is not provided. Instead, information is displayed in an abstracting fashion to be suitable for the non-specialized reader.

Firstly, a general sequence of experiments (Scheme 5.1) performed on live cells from human donor blood can be summarized as follows:

Monocyte-derived dendritic cells (MoDCs) of the appropriate HLA haplotype (HLA A02:01) isolated from donor blood were combined with the respective peptide (as solution, proteoliposome suspension or conjugated to beads) and incubated for the indicated time periods. The unabsorbed antigen was washed from the cell surface and the cells were prepared for detection of MHC-bound epitopes.

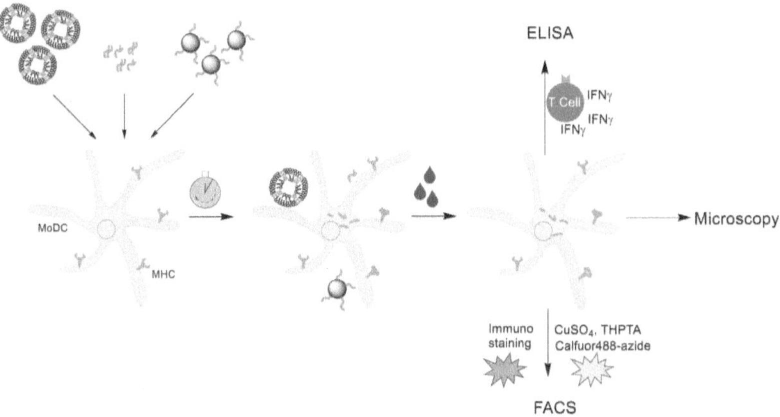

Scheme 5.1: Schematic representation of immunological experiments performed on live dendritic cells.

T cell activation assay

A classic quantification method for presentation or cross-presentation of antigen in live cells is the T cell activation assay. It is executed by addition of cytotoxic T cells that have

---

[3] Radboud Institute for Molecular Life Sciences in Nijmegen and Groningen Biomolecular Sciences and Biotechnology Institute

been transfected with specific receptors which are available for many commonly investigated epitopes. Recognition by a T cell receptor, being selective to epitopes bound to MHC and highly sensitive, causes multiple downstream effects including the synthesis and secretion of pro-inflammatory molecules called cytokines. IFNγ, a homodimeric protein of approximately 40 kDa, is one type of cytokines released by the T cells to activate macrophages as part of the adaptive immune response. IFNγ can be quantified by enzyme-linked immunosorbent assay (ELISA). ELISA a detection method based on specific antibody-antigen interactions (Scheme 5.2). It requires multiple coating and washing steps inside of a well plate during which reaction sites are gradually built up on the surface of the wells. The number of these reaction sites is proportional to the amount of antigen (in this case IFNγ) in the sample. The final coating involves an enzyme which can be quantified by conversion of a substrate which can be detected photometrically. A popular choice for the enzyme is horseradish peroxidase which catalyzes the oxidation of 3,3',5,5'-tetramethylbenzidine. At acidic pH, the resulting diimine has an absorption maximum at 450 nm.

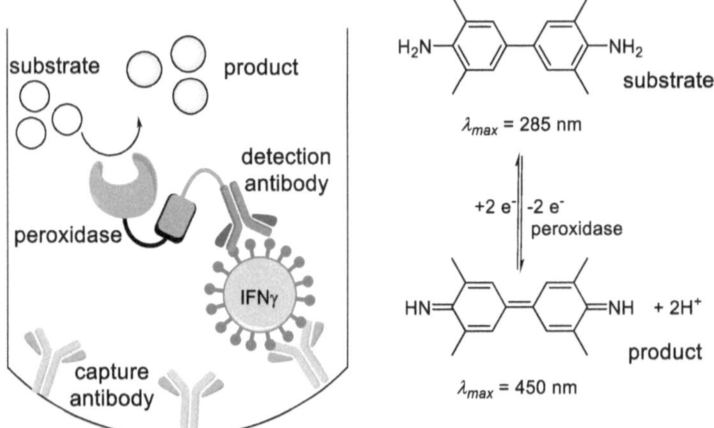

Scheme 5.2: Schematic representation of ELISA used to indirectly quantify presentation and cross-presentation of antigen on the surface of dendritic cells.[216] IFNγ secreted by activated T cells is adsorbed to the surface of a reaction well which was precoated with a capture antibody. A detection antibody binds to the cytokine and by streptavidin biotin interaction horseradish peroxidase is attached. The enzyme catalyzes the conversion 3,3',5,5'-tetramethylbenzidine to the yellow diimine, which can be quantified spectroscopically. Thorough washing steps ensure quantitative correlation of substrate conversion and IFNγ, the total concentration of which can be determined with the help of a calibration series.

Bio-orthogonal labelling and quantification by FACS

Alternatively, a new assay involving bio-orthogonal labeling of the presented epitope was translated for the use on human DCs.[40] An amino acid position inside the epitope which is inessential for binding to MHC and optionally recognition by T cells is exchanged for artificial amino acids propargylglycine or azidohomoalanine which can react in a copper(I)-catalyzed azide-alkyne cycloaddition, also known as click-reaction. Conveniently, quantum yield for fluorescent dye Calfluor488 (Scheme 5.3) is increased by several orders of magnitude upon formation of the triazole,[217] reducing background fluorescence of excess dye significantly. For this assay, cells must be fixed with paraformaldehyde and analyzed by fluorescence-activated cell sorting (FACS) or more broadly referred to as flow cytometry. FACS is a powerful technology used to study cell populations in modern biology laboratories.[218] The underlying principle is fairly simple, however high-throughput real-time analysis of large amounts of data make FACS machines an indispensable tool in immunology research. Cell suspensions are passed through a flow chamber and focused with a sheath fluid to flow one cell at a time through a laser beam (Scheme 5.3 left). Analyzing forward scatter and side scatter caused by each cell, the size and internal complexity can be determined, indicating the identity of the cell. At the same time, fluorescence can be measured, a dichroic mirror setup allowing for simultaneous detection of several fluorophores. Statistical analysis provides information about the type and number of cells present in a sample and correlates the number of cells carrying one or several types of fluorescent markers. Additionally, physical sorting of the cells can be performed by previously defined optical parameters. Modern FACS machines process several thousand cells per second.

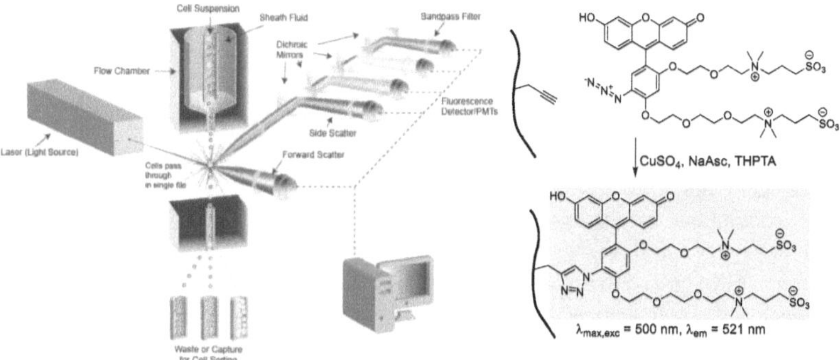

Scheme 5.3: Left: Scheme of flow cytometry and its critical components described in the text.[219] Right: Calfluor 488 before and after copper-catalyzed cycloaddition to MHC bound epitope on the cell surface. The azide and alkyne positions could also be exchanged. The fluorescence quantum yield is increased by a factor of 250 after click reaction.[220]

To account for cells that did not survive the treatment with the artificial antigens, cells were additionally treated with a viability dye eFluor™ 780 (λ_{exc} = 633 nm, λ_{exc} = 780 nm) prior to fixation. The amine reactive dye is able to permeate cellular membranes of dead cells while it can only react with surface bound amines in live cells. Thus, dead cells can be excluded in the FACS analysis by a high intensity of 780 nm fluorescence.

Confocal microscopy

To complement the quantification assays, cells were observed by confocal microscopy, which allows to follow the trafficking of antigen over the course of cellular processing. Reaction with Calfluor488 was not suited for intracellular staining. Therefore, peptides for this purpose were either *N*-terminally conjugated to Atto647 prior to incubation with the cells or marked by intracellular staining with AF568 picolyl azide after incubation with the cells. The latter carries its own copper-chelating motif (Figure 5.1), which increases the local Cu(I) concentration, enhances the efficiency of the click reaction and makes it more biocompatible.[221]

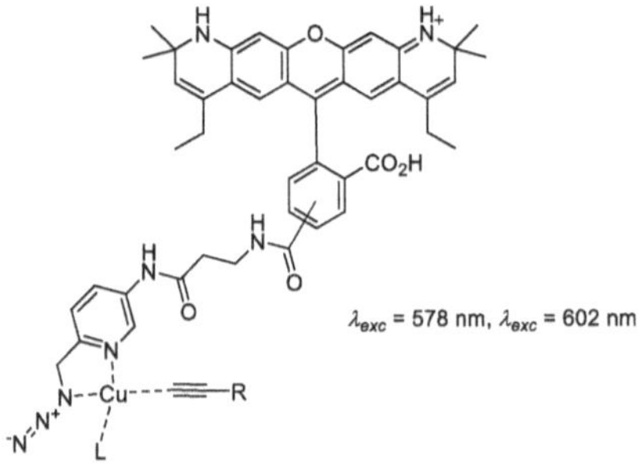

Figure 5.1: Cu(I) complexed by AF568 picolyl azide, alkyne and a click-ligand like THPTA.

5.1 Tracking cross-presentation with bio-orthogonal labeling

Verifying presentation and cross-presentation of antigen on the surface of dendritic cells is a lengthy and labor-intensive process which takes weeks until results can be obtained. Typical immunological assays require additional cell types, cytotoxic T cells, which secrete measurable cytokines upon recognition of epitopes bound to MHC molecules. Extra cells that need to be cultivated, nurtured and treated appropriately for the experiments can be a source of deviations in the obtained results. PAWLAK *et al.* published an assay that can complement conventional measuring methods by eliminating the detour via T-cells.[40] They equipped well established antigen model OVA[257-264] SIINFEKL with bio-orthogonal groups, either via propargylglycine or azidohomoalanine, that could be addressed with copper(I)-catalyzed azide-alkyne cycloaddition. They found that individual positions of the epitope could be exchanged for the unnatural residues and still be bound to MHC I in a mouse cell line. Also, some of the positions were sufficiently exposed to the exterior to be accessible for reaction with Calfluor488 equipped with a moiety for click-reaction, which favorably increases fluorescence quantum yield by a factor of 250 after conjugation.[220] Binding to MHC I could be validated by quantification of Calfluor488 fluorescence on the surface of the cells. This method could be interesting for complementing immunological studies of clinically relevant epitopes when translating it for use on human dendritic cells.

A selection of known epitopes from tumor antigens that would be cross-presented on MHC I in haplotype HLA-A2 dendritic cells was screened for being verifiable both by T cell activation and labeling with Calfluor488-azide (Table 5.1).

*Table 5.1: Peptides synthesized for this chapter. Entries marked with * were obtained from commercial sources.*

Molecule entry	Name	Sequence
50	NY-ESO1_short*	SLLMWITQV
51	NY-ESO1_long*	LQQLSLLMWITQCFL
52	NY-ESO1_short_pra5	SLLM{pra}ITQV
3	NY-ESO1_long_pra5	LQQLSLLM{pra}ITQCFL
53	NY-ESO1_short_pra7*	SLLMWI{pra}QV
54	NY-ESO1_long_pra7*	LQQLSLLMWI{pra}QCFL
55	NY-ESO1_short_pra8*	SLLMWIT{pra}V
56	NY-ESO1_long_pra8	LQQLSLLMWIT{pra}CFL
57	Biotin-NY-ESO1_long	Biotin-aca-LQQLSLLMWIT{pra}CFL
58	Biotin-NY-ESO1_long_pra8	Biotin-aca-LQQLSLLMWIT{pra}CFL
59	Atto647-NY-ESO1_long_pra8	Atto647-LQQLSLLMWIT{pra}CFL
60	gp100_long	VTHTYLEPGPVTAQVVL
61	gp100_short_pra5	YLEP{pra}VTA
62	gp100_long_pra5	VTHTYLEP{pra}VTAQVVL
63	gp100_short_pra7	YLEPGP{pra}TA
64	gp100_long_pra7	VTHTYLEPGP{pra}TAQVVL
65	gp100_short_pra8	YLEPGPV{pra}A
66	gp100_long_pra8	VTHTYLEPGP{pra}TAQVVL
67	Mart1_short_pra5	EAAGI{pra}ILTV
68	Mart1_long_pra5	TTAEEAAGI{pra}ILTVILGV
69	Mart1_short_pra8	EAAGIGI{pra}TV
70	Mart1_long_pra8	TTAEEAAGIGI{pra}TVILGV
71	Mart1_short_pra9	EAAGIGIL{pra}V
72	Mart1_long_pra9	TTAEEAAGIGIL{pra}VILGV

First, applicability in T cell activation assay was tested. This is important to differentiate between epitopes cross-presented on MHC I and residual peptides adhered to the cell surface. In a concentration dependent assessment of NY-ESO1, short peptides modified with {pra} in different positions (Figure 5.2 left) NY-ESO1_short_pra7 (**54**) was shown to be recognized by CD8$^+$ T cells transfected with the corresponding T cell receptor just as well as the native sequence (**50**).

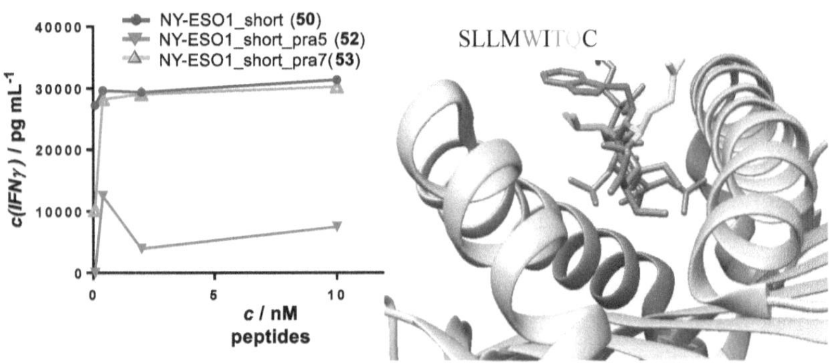

Figure 5.2: Left: Screening for modified epitope positions by T cell activation assay. Right: Crystal structure of NY-ESO1 inside the binding groove of MHC I haplotype HLA A2 (PDB ID: 1S9W).[222] The illustration was prepared with USCF Chimera.

The crystal structure of the epitope inside the binding groove (Figure 5.2 right)[222] unfortunately suggests, that position 7 (threonine) is pointing to the side of the groove and while not essential for T-cell recognition would not be well accessible for labeling. Still, peptides summarized in Table 5.1 were tested for labeling with Calfluor488-azide after incubation with dendritic cells. Cell samples for incubation times of 0 h, 6 h, and 12 h were prepared and combined with peptides at a final peptide concentration of 20 µM. After incubation at 37 °C for the assigned duration, excess peptides were washed from the surface and the cells were fixed with paraformaldehyde. Prior to fixation cells were stained with the viability dye eFluor780 (eBioscience, *Thermofisher scientific*, (Waltham, USA)). Click-reaction with Calfluor488-azide was catalyzed by 1 mM $CuSO_4$ with THPTA as ligand for 2 h. Additionally, live single cells were gated by flowcytometry and the geometric mean of the Calfluor488/FITC channel was calculated.

Analysis of the screening results (Figure 5.3) revealed, that most of the examined artificial antigens cannot be labeled when bound to MHC-I. The highest fluorescence values were obtained for NY-ESO1_short_pra8 (**55**) and NY-ESO1_long_pra8 (**56**), in positive correlation with incubation times. As expected from residue orientation, labeling at position 7 showed only moderate intensities with NY-ESO1_short_pra7 (**53**) with about 10 x lower signal than the NY-ESO1_short_pra8 (**55**). However, it could still be considered when parallel application of T cell activation assay and bio-orthogonal labeling is intended.

Further investigation of model peptide NY-ESO1_long_pra8 was performed by microscopy with *N*-terminally labeled Atto647-NY-ESO1_long_pra8 (**59**). Unfortunately, recordings of cells after incubation with the peptide and presumed removal of excess material by washing showed adhesion of peptides to the cell wall. This is in correlation with the behavior experienced during purification attempts after solid phase synthesis (see chapter 2). Atto647-NY-ESO1_long_pra8 (**59**) could barely be purified, as the bulky dye helped disrupt intramolecular H-bonds of the peptides. NY-ESO1_long_pra8 (**55**) in fact (as other NY-ESO1_long derivatives) aggregated so severely that it would not be fully homogenized by HFIP, appearing dissolved but not passing through a syringe filter, and could therefore not be purified by HPLC.

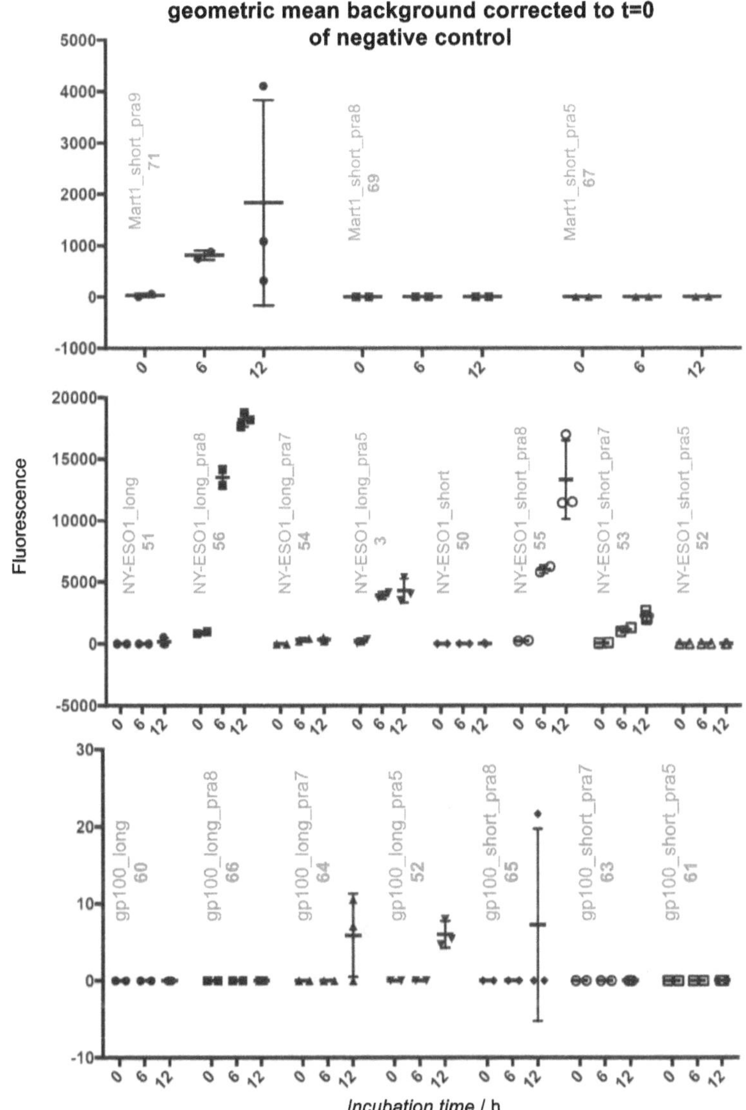

Figure 5.3: Screening for "clickable" epitope positions. Fluorescence plotted against incubation time.

To disable adhesion to the cell surface peptides were equipped with a biotin tag and immobilized on streptavidin coated latex beads. Latex beads constitute a more controlled form of peptide administration and more closely mimic typical fodder for DCs: dead cell material. So far, no cross-presentation for these peptides could be verified. However, exploiting propargylglycine for intracellular click-reaction with AF568-picolyl-azide allowed tracking of the peptides on their way through the cell (Figure 5.4).[221] After 2 h of incubation "clickable" antigen Biotin-NY-ESO1_long_pra8 (**58**) can be localized inside of lysosomes. Beads with non-clickable control peptide Biotin-NY-ESO1_long (**57**) are also recognizable by their characteristic shape but do not react with the fluorophore-azide.

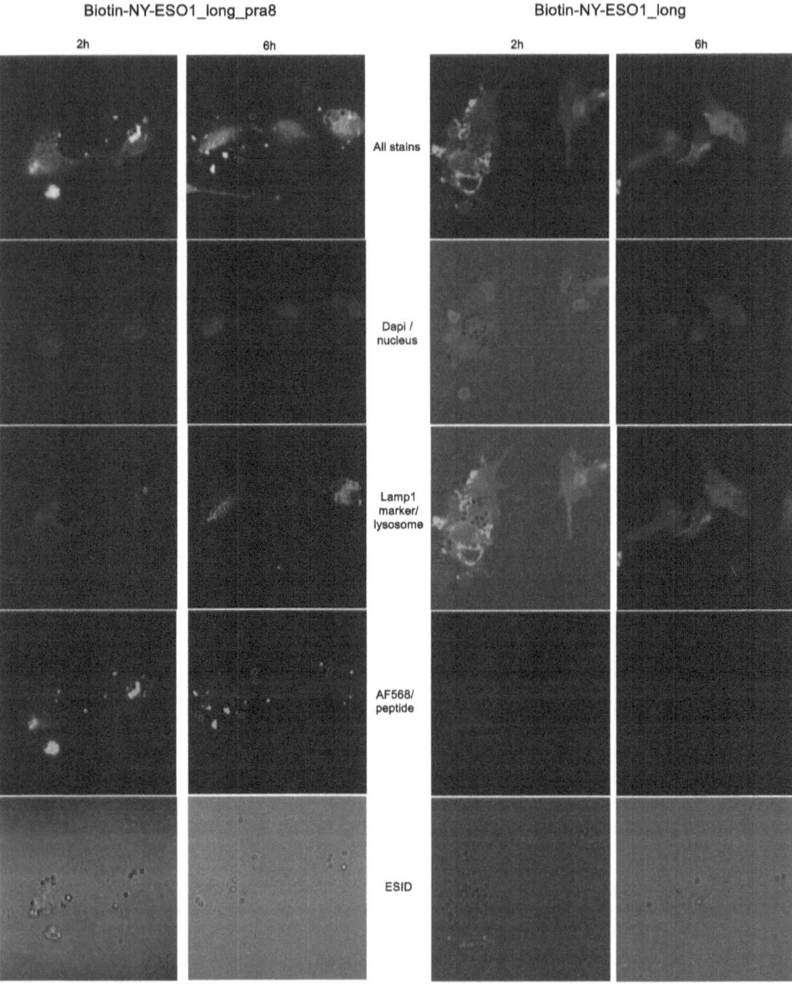

Figure 5.4: Confocal microscopy of dendritic cells treated with biotinylated antigen immobilized on streptavidin coated latex beads after 2 h and 6 h incubation time. Left: Biotin-NY-ESO1_long_pra8 (58) treated cells. Right: Cells treated with control peptide Biotin-NY-ESO1_long (57).

As argued in section 1.2.3, membrane-buried epitopes may play a big role in restricting pathogens from evading adaptive immune responses by mutation. However, mechanisms for processing TMHs in cross-presentation have not yet been described. Therefore, experiments were designed to model antigens embedded inside of lipid membranes. The generalized procedure as shown in Scheme 5.1 was carried out with MoDCs of haplotype HLA-A02:01. Transmembrane peptides were administered as proteoliposomes obtained from extrusion of rehydrated peptide/lipid films (DOPC/DOPS 70:30, 1:80 peptide to lipid ratio). After incubation overnight, excess proteoliposomes were removed by washing with buffer and the cells were combined with naïve CD8$^+$ T-cells transfected with the receptor for NY-ESO1[157-165]. After 72 h, concentration of IFNγ was quantified by ELISA.

The sequences of the used transmembrane peptides were rationalized as follows: Known tumor antigen NY-ESO1[157-165] was flanked with few amino acids of the native sequence on each side and then hydrophobic amino acids allowing to form a transmembrane helix and to contain tryptophan in the N- and C-terminal regions to secure them inside the lipid bilayer. Also, the epitopes should contain a bio-orthogonal group, alkyne or azide, for future labeling with a fluorescent dye. The positioning of epitope inside the TM was varied and the flanking amino acids were adapted to predict an α-helical TM peptide and to be anchored. As control, a section of syntaxin-3 transmembrane domain was equipped with a {pra} position and edited to match the artificial TMPs in length and flanking amino acids. Peptides prepared for this chapter are summarized in Table 5.2, indicating peptides that were synthesized for future experiments but not yet used on grey background. Peptides marked with an asterisk (*) were obtained from commercial sources.

Table 5.2: Overview of peptides synthesized for this chapter, epitopes are highlighted.

Molecule entry	Name	Sequence
53	NY-ESO1_short_pra7*	SLLMWI{pra}QV
54	NY-ESO1_long_pra7*	LQQLSLLMWI{pra}QCFL
73	TM1	AAAWCLQQLSLLMWITQCFLPVFLAWAAA
74	TM1_pra5	AAAWCLQQLSLLM{pra}ITQCFLPVFLAWAAA
75	TM1_azi7	AAAWCLQQLSLLMWI{az}QCFLPVFLAWAAA
76	TM1_pra7	AAAWCLQQLSLLMWI{pra}QCFLPVFLAWAAA
77	TM1_pra8	AAAWCLQQLSLLMWIT{pra}CFLPVFLAWAAA
78	TM7_pra7	AAAWLLCLVVLSLLMWI{pra}QCFLPVFWAAA
79	TM8_pra7	AAAWFVLLCLVVLSLLMWI{pra}QCFPWAA
80	TM9	AAAWPFVLLCLQQLSLLMWITQCFLWAAA
81	TM9_pra5	AAAWPFVLLCLQQLSLLM{pra}ITQCFLWAAA
82	TM9_pra7	AAAWPFVLLCLVVLSLLMWI{pra}QCFLWAAA
83	TM9_azi7	AAAWPFVLLCLVVLSLLMWI{az}QCFLWAAA
1	TM9_pra8	AAAWPFVLLCLQQLSLLMWIT{pra}CFLWAAA

84	ControlWW	AAAWLIIIIVIVVVLLGI{pra}LALIIGLWAAA
85	ICP47	MSWALEMADTFLDTMRVGPRTYADVRDEINKRGRE

Selected peptides were tested for their secondary structure inside of lipid membranes by CD spectrometry. Peptide/lipid films were rehydrated and homogenized by extrusion through polycarbonate membranes (Ø = 100 nm) to produce proteoliposomes (2 mg/mL lipids (70% DOPC, 30% DOPS), 0.1 mg/mL peptide, 10 mM phosphate buffer pH 7). CD signals measured between 190 and 250 nm are plotted in Figure 5.5. With minima at around 210 nm and 222 nm, the curves represent the characteristic shape produced by an α-helix. ControlWW (**84**) and TM1 variants (**75** and **76**) produce overall more intense signals than TM9 variants (**82** and **83**).

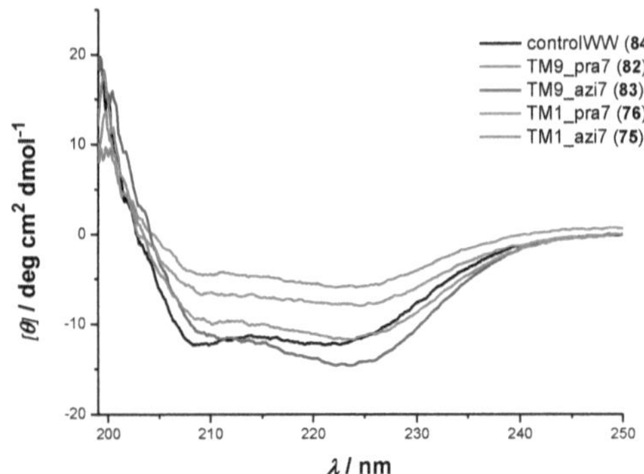

Figure 5.5: Secondary structure of artificial transmembrane peptides inside of liposomes was verified by CD spectroscopy. Proteoliposome suspensions with a peptide to lipid ratio of 1:80 and a final peptide concentration of 0.1 mg/mL were measured in 10 mM phosphate buffer at pH 7.

Before any mechanistic investigations could take place, a functioning model TM peptide needed to be found that would be effectively cross-presented in undisturbed cells. Proteoliposomes were added to the MoDCs at different concentrations between 0.1 µM and 10 µM final peptide concentrations and removed after overnight incubation. As a comparison, MoDCs were treated with soluble peptides. Figure 5.6 depicts the relations between IFNγ secretion and the applied peptide concentration.

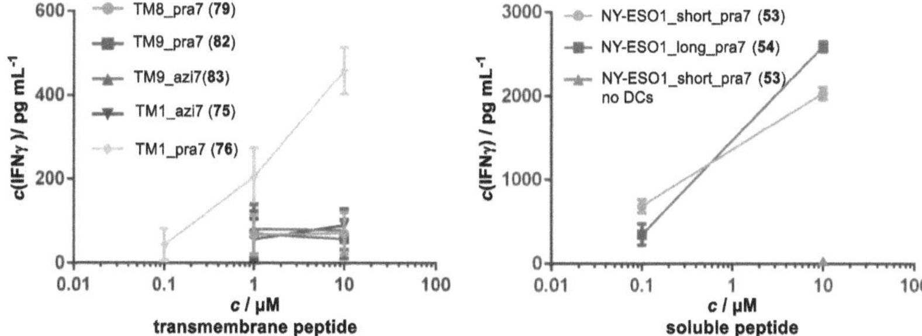

Figure 5.6: Concentration dependent cross-presentation of artificial peptides. Left: Transmembrane peptides were administered as proteoliposome suspension. Right: Soluble peptides were added as solution.

A dose-dependent TCR response is distinctly measurable for TM1_pra7 (**76**) while the other TMPs do not show any measurable effect. Cross-presentation of soluble peptides NY-ESO1_short_pra7 (**53**) which does not need processing to be bound to MHC-I and NY-ESO1_long_pra7 (**54**) that does require trimming before being loaded on the antigen presenting complex likewise shows a dose-response to applied peptide. Interestingly, the overall response to soluble peptide is about fivefold compared to the TMP. Nonetheless, TM1_pra7 (**76**) and the corresponding soluble analogs were deemed as useful tool to compare mechanisms of cross-presentation of soluble and membrane buried antigens.

As mentioned in section 1.2.2, proteasome and TAP are good targets to differentiate between the two cross-presentation pathways that are most prominently discussed to date. MG132 is a potent synthetic proteasome inhibitor from the family of peptide aldehydes.[223] It was administered to cell populations that would be treated with the three previously identified model peptides at different concentrations (Figure 5.7). While high concentrations >0.1 µM of inhibitor caused cell death in all populations, at 0.1 µM a difference in response by the different populations could be noted. Cross-presentation of soluble antigens, both short and long variant were reduced in comparison to uninhibited cells whereas cross-presentation of membrane-buried antigen remained undisturbed.

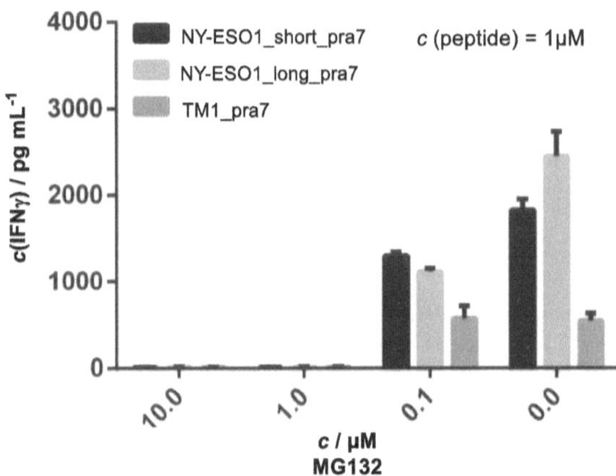

Figure 5.7: Influence of proteasome inhibitor MG132 on cross-presentation of soluble and membrane-buried antigen.

This could be a first hint to TM protein-derived antigens not being cross-presented by the P2C pathway. To supplement this finding future control experiments will be performed with TM1 which does not carry modifications in the epitope. Furthermore, it is planned to use TAP inhibitor ICP47 (**85**) as next step as proteasome inhibition alone is not sufficient to distinguish if membrane proteins are presented via P2C or the vacuolar pathway.

5.3 Presentation on MHC-II verified by bio-orthogonal labeling

In contrast to MHC-I molecules, which can be found on the surface of all types of cells in the human body, MHC-II is only expressed in specialized antigen presenting cells like dendritic cells. While their function is similar, the MHC-II presented antigens usually originate from outside of the organism, like from bacteria or viruses, and are processed to epitopes of 14 amino acids length. After uptake of foreign material early endosomes and phagosomes must fuse with other compartments, recruit cytosolic proteins to their membrane, and transform to lysosomes by a process called maturation. Lysosomes have a lower pH than their predecessors which is achieved by protein pumps and activates proteases such as cathepsin-S breakdown of the ingested material. Appropriately sized peptides are loaded on MHC-II and transferred to the cell surface to be recognized by CD4+ T-cells and trigger adaptive immune responses.[159]

Maturation of early endosomes is accompanied by a change in composition of phosphoinositide lipids in their membrane as a marker for the change in organellar identity (Scheme 5.4).[224] Conversion of phosphatidylinositol-4,5-bisphosphate (PI(4,5)P_2) to phosphatidylinositol 3-phosphate (PI(3)P) can be accomplished by various kinases and phosphatases, but for the progression to phosphatidylinositol 3,5-bisphosphate (PI(3,5)P_2), phosphoinositide 5-kinase (PIKfyve) is the only known catalyst.[225,224] Despite the plausible link of PI(3,5)P2 and PIKfyve with antigen presentation on MHC-II, a direct connection had not been shown previously.

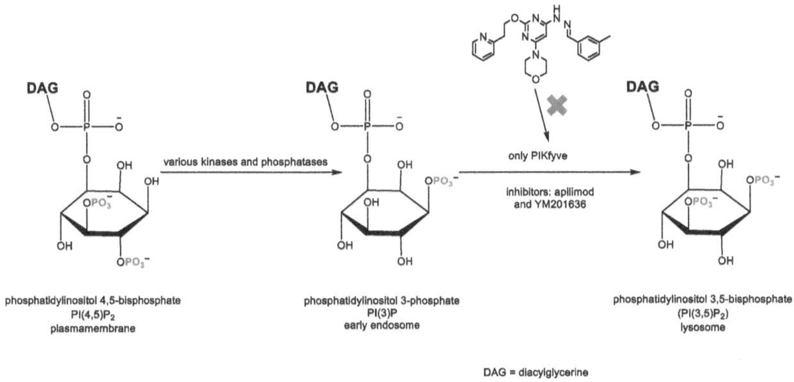

Scheme 5.4: Conversion of phosphoinositide lipids in the course of early endosome maturation to lysosomes.

The influence of PIKfyve on MHC-II presentation was tested with the help of apilimod and YM201636, both selective inhibitors of PIKfyve.[41,226,227] In donor blood derived dendritic cells, presentation of OVA[323-339] with and without inhibitors was compared by T cell activation assay. A reduction of ~25% in IFNγ excretion was observed over time. However, long exposure time to the inhibitors also reduced viability in both DCs and CD4$^+$ T-cells and could distort the results. As a complementary and more direct analysis method, the bio-orthogonal labeling assay[40] described in previous chapters was adapted for MHC-II presentation in human dendritic cells. As a model antigen, epitope of hemagglutinin HA[322-334] from influenza A virus was extended by four native amino acids on both sides (HA[318-338]). Position K326 was chosen for exchange with {pra}. Model antigen HA_long_clickable (**86**) and control peptides HA_short non-clickable (**87**), HA_short_clickable (**88**) and HA_long_non-clickable (**89**) were synthesized (Table 5.3). Short control peptides were chosen consisting only of the epitope sequence. Only peptides that are processed inside the lysosome and need to be trimmed to size are actively loaded to MHC-II. Thus, the short peptides were not expected to be presented on the cell surface.

Table 5.3: Overview of peptides used for this chapter.

Molecule entry	Name	Sequence
87	HA_short non-clickable	PKYVKQNTLKLAT
88	HA_short_clickable	PKYV{pra}QNTLKLAT
89	HA_long_non-clickable	YGACPKYVKQNTLKLATGMRN
86	HA_long_clickable	YGACPKYV{pra}QNTLKLATGMRN

First, model antigen HA_long_clickable (**86**) was tested for being accessible for labeling with Calfluor488-azide. Solutions of antigen and control peptides were added to HLA-

DR1 DCs and incubated for 1 h, 2 h, 3 h, 4 h, 5 h, and 24 h. Excess peptide was removed by washing and the cells were fixed with paraformaldehyde. Click-reaction with CalFluor488-azide was performed and MHC-II molecules were additionally marked with immunostaining. Cells were analyzed by flow cytometry. CalFluor488 fluorescence of HA_long_clickable (**86**) was found to increase over the course of five hours and drop at the 24 h datapoint (Figure 5.8, left). The drop could be explained by internalization and degradation of MHC-II molecules with the bound epitopes by the cells. In contrast, the fluorescence in all control peptide treated cell populations stayed at a constant background level.

For testing the influence of PIKfyve inhibition on MHC-II presentation, antigen solution was added to HLA-DR1 DCs and incubated for 2 h. Inhibitors apilimod or YM201636, or DMSO as control were added and incubated with the cells for 3 h before washing, fixing and labeling with CalFluor488. Instead of adding inhibitor, trypsin could be added to remove epitopes from MHC-II after five total hours of incubation with the antigen. Fluorescence was analyzed by FACS. Depending on the donor, fluorescence was reduced by 20-80% through inhibition of PIKfyve with apilimod and YM201636 (Figure 5.8 right). For apilimod the effect was less pronounced (20-40% reduction) but more closely distributed over the donors.

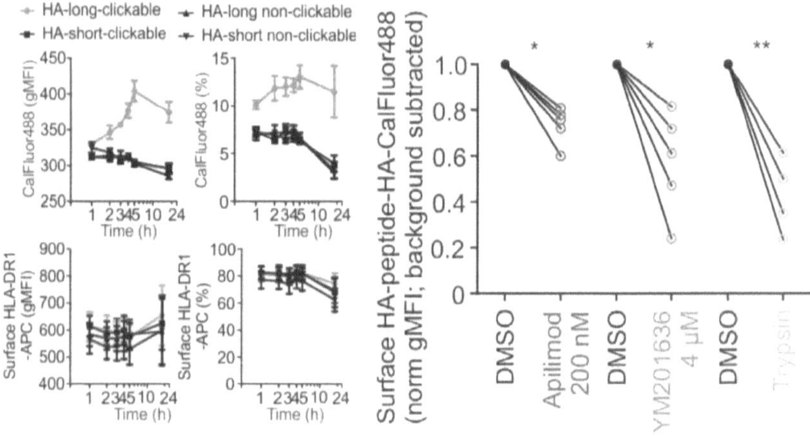

Figure 5.8:Left: Testing of model antigen for being accessible for biorthogonal labeling. Right: Fluorescence after treatment of DCs with inhibitors. Figures were published in iScience **2019**, *11, 160.*[141]

Overall, bio-orthogonal labeling was established as complementary assay to study MHC-II presentation. A direct influence of PIKfyve inhibition on MHC-II presentation previously found by T-cell activation assay could be confirmed by bio-orthogonal labeling of epitope bound to MHC-II.

In this chapter we contributed artificially modified peptides to a rather ambitious project in immunology research. Two main aspects were in the center of this project: the question of how membrane buried epitopes are cross-presented in dendritic cells and the aim to study it with a bio-orthogonal labeling strategy as a complementary method to established but laborious immunological assays.

In section 5.1 the translation of the method previously reported by PAWLAK *et. al.*[40] for mouse models to the investigation of clinically relevant soluble epitopes in human cells was described. Screening known tumor antigens for suitable residues to exchange for bio-orthogonal alkyne linkers for labeling with "clickable" fluorophore Calfluo488 azide provided one promising hit with NY-ESO1_pra8 in the long and short version. For these peptides, T cell activation could not be verified, so true cross-presentation could not yet be confirmed. On the other hand, in NY-ESO1_pra7, T cell activation was in the range of the native sequence, while labeling was about 10 times less efficient than in the best hit. With all peptides derived from the NY-ESO1 antigen severe adhesion to the cell surface complicated the evaluation of the labeling results. The aim to label the antigen inside of the MHC-I binding groove was abandoned for the time being and the alkyne linker was used for intracellular staining of antigen conjugated to latex beads by biotin/streptavidin interactions. The stained peptides on beads could be observed by confocal microscopy in different compartments of the cells over the course of cellular processing and will be used as a tool to track antigen in future applications.

While the linker (alkyne or azide) for bio-orthogonal labeling was integrated in the membrane buried epitopes investigated in section 5.2 for future analysis, cross-presentation was primarily verified by T cell activation. Artificial transmembrane peptides were designed and synthesized with the NY-ESO1 epitope integrated in different positions of the sequence. One of the constructs was verified on MHC-I by the classic immunological assay and could be used to explore the yet unknown processing pathways of membrane buried epitopes. Inhibition with GM132 suggested that TMP processing is independent of proteasomes, indicating that TM proteins do not take the phagosome-to-cytosol pathway. This finding will be verified by inhibition of TAP with ICP47 in upcoming experiments.

Finally, bio-orthogonal labeling found a relevant application in a related project described in section 5.3, verifying presentation of antigen on MHC-II. PIKfyve, an enzyme which catalyzes the conversion of membrane lipid PI(3)P to PI(3,5)P$_2$, had been shown to be indispensable for lysosome maturation and subsequent antigen presentation on MHC-II by T cell activation. However, the results were questioned because of the adverse effect of the used inhibitors on T cell viability. Thus, the T cell independent labeling assay was used to provide complementary information. In the used model antigen derived from hemagglutinin from influenza A virus no conflict was found between accessibility for bio-orthogonal labeling and recognition by the T cell receptor. Inhibition with pharmaceuticals apilimod and YM201636 for different durations and correlation with labeling efficiency of click-reaction with Calfluor488 azide confirmed the direct influence of PIKfyve on antigen presentation on MHC-II. The results could be published as part of a more extensive study of phagosome maturation.[41]

6 Summary and outlook

In this book, artificially modified peptides were synthesized to study SNARE-mediated membrane fusion *in vitro* and antigen processing by antigen presenting cells *ex vivo*. Apart from vesicle trafficking being part of both processes, the projects were connected by the challenging peptide sequences – in parts buried inside lipid membranes – that needed to be provided in good purities to obtain reliable results. Optimizing and streamlining the handling of aggregation prone peptides in solid phase peptide synthesis by Fmoc strategy and RP-HPLC purification was a central task of this work and could be successfully accomplished for most peptides. Treatment of HPLC samples with HFIP proved to be a crucial step so that peptide aggregates could be monomerized prior to purification. With proper pretreatment of the samples, only minor changes to standard HPLC procedures were needed to isolate the SPPS products.

For the SNARE project, a set of fully peptidic model fusogens developed in the DIEDERICHSEN group by KARSTEN MEYENBERG was to be refined with photocleavable protecting groups to provide temporal control over the fusion of lipid vesicles. The E3Syb/K3Sx fusion pair was previously designed to closely mimic SNARE mediated fusion by exchanging the SNARE motif for heterodimeric parallel coiled coils E3 and K3 and conserving the linker and transmembrane domain of synaptic SNAREs syntaxin-1A and synaptobrevin-2.[26] The aim of this work was to develop a photoprotection strategy that would reversibly halt the vesicles in intermediate stages preceding full fusion – docking and hemifusion – to study different factors important to the transition between those stages. The caging should be effective in preventing lipid mixing but at the same time minimally intrusive to still allow interaction of the recognition units for docking of the reconstituted vesicles. Furthermore, the activity of the fusogens was to be recovered fast by uncaging with UV irradiation so that time resolved measurements could provide high information value.

Disturbing coiled coil interactions in a precise manner was challenging because the tightly packed hydrophobic core formed by the peptide pair is mostly stabilized by non-selective hydrophobic interactions. Furthermore, the amino acids in question – leucin and isoleucine – cannot be covalently protected in a reversible way. Instead, the flanking glutamic acids and lysins were used to introduce steric hindrance close to the hydrophobic core. Literature known caging groups NVOC, DMNPB and DEACM were used to synthesize a selection of E3Syb and K3Sx derivatives protected with one or two caging groups in the membrane proximal heptads of the recognition units. Combinations of complementary peptide pairs were studied in a total lipid mixing assay based on FRET. The assay was useful to determine if interaction of the recognition pairs could be suppressed and if activity could be recovered in comparison to the positive control E3Syb/K3Sx. The combination of E3(DMNPB)$_2$Syb and K3(NVOC)$_2$Sx (**28** + **27**) showed a promising inhibitory effect on the fusogenicity of the SNARE mimetics and lipid mixing could be verified after irradiation with 347-400 nm. However, long irradiation times (2 min) required to uncage the peptides negatively impacted the assay by photobleaching and restricted access to fluorescence data at the onset of lipid mixing. Efforts to remedy this drawback by the more efficiently released DEACM were cut off by the synthesis failure of E3(DEACM)$_2$Syb (**30**) most likely by pyroglutamate formation.

Alternatively, a novel protection strategy was developed which would place bulky and photolysis-efficient DEACM in between the coiled coils, thus more precisely targeting the hydrophobic interactions. Inspired by stapled peptides intramolecularly linked by alkene metathesis, DEACM was equipped an allyl linker which could connect two caged lysine or glutamic acid residues spanning two DEACM groups over the relevant isoleucine and leucine positions. Photolysis kinetics and the produced side products were studied with the help of soluble derivatives of the SPPS building blocks dimerized by metathesis. Two stapled E3 derivatives were synthesized to establish a synthesis strategy and to examine the structural impact of the linkers. While an influence on coiled coil interactions remains to be confirmed, an induction of α-helical secondary structure of the otherwise unstructured isolated E3 could be verified by CD spectroscopy in peptide **32**. Transferred to a stapled SNARE mimetic, this could be beneficial to promote docking of vesicles by preorganization of the recognition unit.

Synthesis and purification of stapled K3(DEACM)$_2$^{butenyl}Sx (**49**) could be achieved implementing lessons learned in the synthesis of the stapled test-peptides and purification of other SNARE mimetics. LC-MS analysis proved useful in identifying product containing fractions obtained from HPLC separation. In combination with non-caged E3Syb, the pair is now ready to be tested as minimally perturbed caged fusion pair. Following a total lipid mixing assay, the exact fusion states before and after irradiation will be distinguished in future experiments with the help of an inner lipid mixing assay and fluorescence correlation spectroscopy (FCCS). FCCS can verify docked vesicles which are not visible in lipid mixing assays by spatially correlating fluorophores now placed in both vesicle populations.[228] Content mixing – exploiting self-quenching of encapsulated fluorophores such as fluorescein derivatives which is relieved upon fusion with empty vesicles – may help verify complete fusion of vesicles after uncaging. The caged fusion pair will be useful to test the influence of lipid composition on the model membrane and comprehend the role of specific amino acids of linker and transmembrane domains on the transition between fusion states.

In collaboration with the VAN DEN BOGAART group processing of model antigens was investigated in antigen presenting cells derived from donor blood. A selection of artificial transmembrane peptides was designed to incorporate known tumor epitope NY-ESO1 and obtained from SPPS and RP-HPLC in good purity. Bio-orthogonal azide and alkyne moieties were shown to not interfere with cellular processing and were included in the peptide sequences for future fluorescent labeling by copper-catalyzed cycloaddition. In a pioneering investigation, cross-presentation on MHC-I of dendritic cells could be verified for membrane buried epitopes by a classic T cell activation assay. The functional model system was used to follow the pathway by which the antigen was processed to be presented on MHC-I. Administration of selective proteasomal inhibitor GM132 showed that cross-presentation of the transmembrane peptide was not influenced by the protein degrading machinery in contrast to soluble versions of the same epitope. One of the major known processing pathways is the phagosome-to-cytosol (P2C) pathway which is dependent on proteasomal degradation and TAP. Not needing proteasomes for cross-presentation could indicate that membrane buried epitopes are processed by a different pathway than P2C. This finding will be complemented by TAP inhibition in future experiments.

A novel assay based on bio-orthogonal labeling of epitopes bound to MHC was to be translated to study clinically relevant epitopes in human antigen presenting cells.[40] Fluorescent labeling inside the binding grove of MHC would circumvent the need for additional cell types and provide a quick and easy readout for pharmaceutical manipulations and related investigations. Regarding cross-presentation on MHC-I HLA-A2 the chosen epitopes proved to be inadequate to verify processed epitopes on the cell surface due to inaccessibility of the labeling site and aggregation. Instead, the modified epitopes were conjugated to latex beads and tracked by intracellular staining and confocal microscopy (Figure 6.1). On the other hand, click-reaction following cellular processing was successful for an antigen presented on MHC-II. The method was applied as a valuable supplementary measurement investigating the role of phosphoinositide kinase PIKfyve for MHC-II antigen-presentation. A direct link between PIKfyve inhibition reduced antigen presentation was demonstrated by T cell activation assay and could be confirmed by quantifying bio-orthogonal fluorescent labeling of the presenting epitope.[41]

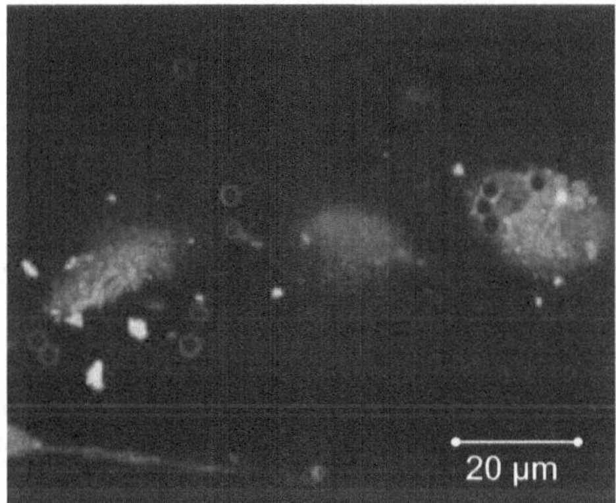

Figure 6.1: Confocal microscopy of dendritic cells processing antigen conjucated to latex beads. The labeled peptides are clearly visible as orange rings.

7 Experimental section

7.1 General

7.1.1 Reagents and solvents

All reagents were purchased in the highest purity available and used as supplied. Amino acid derivatives and coupling reagents were obtained from GL Biochem (Shanghai, China), Iris Biotech (Marktredwitz, Germany), and Roth (Karlsruhe, Germany). Resins for SPPS were aquired from Novabiochem (Darmstadt, Germany) and Sigma Aldrich (St. Louis, USA). Unlabeled and labeled lipids were purchased from Avanti (Alabaster, USA). All other Chemicals were obtained from varying commercial sources. Solvents for reactions were of the grade *pro analysi* (p.a.) or absolute. Technical grade solvents for flash chromatography were distilled prior to use. Solvents for HPLC and spectroscopic applications were of HPLC grade while LC-MS was conducted with LC-MS grade solvents, including LC-MS grade water. Demineralized water was further purified using either a Simplicity water purification system by Merck Millipore (Billerica, USA) or an arium mini lab water system by Sartorius (Göttingen, Germany) to obtain ultrapure water.

7.1.2 Reactions

Air and moisture sensitive reactions were conducted under nitrogen or argon atmosphere in dry solvent. Glassware was therefore heated with a heat gun under reduced pressure and purged with the inert gas using SCHLENK-type techniques. If indicated, light-sensitive compounds were protected from exposure to light by wrapping the glassware with aluminum foil.

7.1.3 Lyophilization

Aqueous or dioxane solutions were frozen in liquid nitrogen and freeze-dried using a Christ Alpha 2-4 LD plus lyophilizer (Osterode im Harz, Germany) connected to a high vacuum pump. In fractions collected from the HPLC the organic content was reduced to a minimum amount with a flow of nitrogen and they were likewise freeze-dried. For small volumes (< 2 mL) a Christ RVC 2-18 CD plus vacuum centrifuge (Osterode im Harz, Germany) connected to the lyophilizer was used. Small volumes (< 2 mL) of poorly soluble transmembrane peptides were lyophilized from neat HFIP.

7.1.4 Storage

Chemicals and resins for SPPS were stored according to suppliers' recommendations at -18 °C, 2 °C or room temperature and under argon when appropriate. Synthesized compounds and cleaved and purified peptides were stored at -18 °C while on-resin-peptides and compounds meant for the determination of extinction coefficients were kept in a desiccator. Light-sensitive compounds were protected with aluminum foil.

7.1.5 Uncaging

Method a) Photolytic cleavage of NVOC and DMNPB was performed in a UV cuvette (1 cm) using an Ark Lamp Source (*66924*) from *Newport*, consisting of a 1000 W *Arc Housing-IGN F/1.0* (*66921*), a 1000 W Hg-Xe lamp (*6295NS*), a power supply (*69920*), a dichroic mirror (280-400 nm, *66226*), and a 347 nm longwave cut-on filter (*20CGA-345*). Samples were irradiated at 1000 W for 10 s to 10 min. Method b) Coumarin based protecting groups were usually cleaved in a stirrable UV cuvette (1 cm) placed on a magnetic stirrer using a 400-415 nm laser (75 mW) from *Laser 2000* (Wessling, Germany). Method c) During total lipid mixing experiments of coumarin protected proteoliposomes a 405 nm (100 mW) laser pointer distributed by *starklasers.com* (London, UK) was directed to the top opening of the fluorescence cuvette inside the fluorospectrometer.

7.2 Characterization

7.2.1 Nuclear Magnetic Resonance Spectroscopy (NMR)

^{1}H- and ^{13}C-NMR spectra were recorded on a *Varian* (Palo Alto, USA) instrument (*Mercury-VX 300, VNMRS-300* or *INOVA-500*). The measurement frequency and the utilized deuterated solvent are indicated with each compound. Residual proton signals of the solvents served as an internal standard. Chemical shifts δ are given in parts per million (ppm) and coupling constants ^nJ are given in Hertz (Hz), where n indicates the order of coupling. Multiplicities are described with the following abbreviations: s (singlet), d (doublet), t (triplet), q (quartet) or m (multiplet). Signals were partially attributed with the help of COSY- (Correlated Spectroscopy), HSQC- (Heteronuclear Single Quantum Coherence) and HMBC- (Heteronuclear Multiple Bond Correlation) spectra.

7.2.2 Mass Spectrometry

All compounds were characterized by electron spray ionization mass spectrometry (ESI-MS) and high resolution ESI (HR-MS) on a maXis or MicroTOF spectrometer from Bruker (Bremen Germany). Values are given in *m/z*.

7.2.3 UV-Vis

A *nanodrop ND-2000c* spectrometer from *Thermo Fischer Scientific* (Waltham, USA) with a *Quartz SUPRASIL QS* cuvette of 1.0 cm path length was used for all UV/Vis related measurements. The LAMBERT-BEER law was applied:

$$c = \frac{A}{\varepsilon \cdot d}$$

with c being the compound concentration (M), A the absorption of the sample, ε the molar extinction coefficient (M^{-1} cm^{-1}) and d the path length (cm).

Molar extinction coefficients of new compounds were measured from solutions with known concentrations. For each determined value, three separate aliquots of about 5 mg were lyophilized inside of Eppendorf tubes and weighed immediately after removal from the lyophilizer because the compounds were very hygroscopic. Dilution series in MilliQ and MeCN were prepared with three independent samples for each concentration and from each aliquot. Concentrations were calculated assuming the molar weight of the neutral compounds as there was no evidence of TFA salts (that could have formed upon HPLC purification) in ^{13}C-NMR. The averaged measured absorptions were plotted against the concentrations. The extinction coefficient was obtained as the slope of the linear fit with the y-intercept set to 0. The plots can be seen in the appendix Figure-A 8 and Figure-A 9.

Peptide concentrations were calculated from absorptions and literature known or measured molar extinction coefficients. For peptides with more than one chromophore the ε at the corresponding wavelength was calculated by summation.

	Absorption wavelength λ [nm]	Extinction coefficient ε [M^{-1} cm^{-1}]	Reference
Tryptophan	280	5690	[229]
Tyrosine	280	1280	[229]
Cysteine	280	120	[229]
NVOC	350	5485	[230]
DMNPB	350	4500	[184]
DEACM	385	16000	[194]
DEACM^{allyl}	380	17018 in MeCN, 17155 in H$_2$O	measured
(DEACM)$_2$^{butenyl}	380	25972 in MeCN, 23121 in H$_2$O	measured

7.2.4 Fluorescence

Fluorescence measurements were performed in a *FP-6200* spectrofluorometer from *JASCO* (Tokyo, Japan). The temperature was controlled by an ETC-272 peltier thermostat by JASCO connected to a Thermo Haake WKL26 water recirculatory by *Thermo Electron Corp.* (Waltham, USA). The samples were placed in a quartz glass cuvette with a light path of 10 x 4 mm (excitation 10 mm, emission 4 mm) from *Hellma Analytics* (Müllheim, Germany). Data was acquired and analyzed with the Spectra Manager software by *JASCO*. In lipid mixing experiments, temperature was adjusted to 25 °C and stirring was set to 900 rpm. Fluorophores were excited at 460 nm with a bandwidth of 5 nm and emission was measured with 5 nm bandwidth. Sensitivity was set to "high" with a data pitch of 1 nm. For time resolved measurements, emission at 530 nm was followed over 1200 s to 7200s. Data was collected every 1 s in 1200 s measurements and every 10 s in longer measurements.

7.2.5 Circular dichroism (CD) spectroscopy

CD spectroscopic measurements were carried out on a J-1500 CD spectrometer from Jasco (Tokyo, Japan) connected to a F250 recirculation cooler from Julabo (Seelbach,

Germany) under continuous flushing with nitrogen. Samples were analyzed in a Quartz SUPRASIL QS cuvette with 0.1 cm pathlength from Hellma Analytics (Mühlheim, Germany). Individual spectra were recorded at 20 °C in a wavelength range of 190-260 nm with a bandwidth of 1 nm, response time of 1.0 s, data pitch of 1 nm and scanning speed of 100 nm/min in "continuous mode". Data was accumulated over 10 measurements. Temperature dependent measurements were performed between 5 °C and 95 °C and followed by the CD signal at 222 nm. Data was collected at each full °C and measurement was started when the temperature stayed within ± 0.20 °C for 10 s.

For better comparability of peptides with different numbers of chromophores (amide bonds) measured, the CD signal was converted to mean residue ellipticity [θ] in deg cm^2 dmol^{-1} by the following formula:

$$[\theta] = \frac{\theta_{obs} \cdot MRW}{10 \cdot d \cdot c}$$

where θ_{obs} is the observed ellipticity (deg), MRW is the mean residue molar weight (g mol^{-1}) calculated from the molar weight of the peptide divided by the number of amide bonds, d is the pathlength (cm) and c is the concentration in (g mL^{-1}).

7.3 Chromatography

7.3.1 Thin Layer Chromatography

The progression of reactions was monitored by thin layer chromatography on *Merck* (Darmstadt, Germany) silica gel 60 F254 coated aluminum sheets. Spots were detected by fluorescence quenching at 254 nm or fluorescence at 366 nm or visualized by TLC staining solutions and heating with a heat gun. Amines were identified with ninhydrin solution (1.5 g ninhydrin and 3 mL acetic acid in 100 mL n-butanol) while a cerium ammonium molybdate (CAM) solution (2.0 g Cer(IV)SO$_4$ and 5.0 g (NH$_4$)$_2$MoO$_4$ in 200 mL 10% aq. H$_2$SO4) was used as a universal stain.

7.3.2 Flash Chromatography

Purification by flash chromatography was performed using *Merck* (Darmstadt, Germany) silica gel type 60 (particle size 40-63 µm) and a pressure of 0.1-1 bar. Silica gel was suspended in the eluent of choice and packed in an appropriately sized glass column equipped with a glass frit. Crude samples were loaded in a thin layer preabsorbed to the fivefold amount by weight of silica gel.

7.3.3 Reversed Phase High Performance Liquid Chromatography (RP-HPLC)

Analytic RP-HPLC of soluble compounds was carried out on a UPLC system *UltiMate 3000* from *Thermo Fischer Scientific* (Waltham, USA) consisting of autosampler-, pumps-, column oven-, detector, and diode array modules of the *3000* series. Compounds that required eluent systems VI, VII and VIII (see below) were analyzed on a *JASCO* (Tokyo, Japan) system with two pump modules (*PU-2080 Plus*), a photo diode array detector (*MD-4015*), and a degasser (*DG-2080-53*). Semi-preparative and preparative

HPLC purification was carried out on either a *ÄKTA basic 10* system by *Amersham Pharmacia Biotech* (Umeå, Sweden), equipped with a pump module (*P-900*) and a UV detection module (*UV-900*) or a *JASCO* (Tokyo, Japan) system consisting of two pump modules (*PU 2080 Plus*), degasser (*DG-2080-53*) and a multi wavelength detector (*MD-2010 Plus*). If needed, a fraction collector (*CHF122SC*) from *Advantec* (Multiplas, USA) was coupled to the *JASCO* (Tokyo, Japan) system by a connector module (*FC-2088-30*). If required, the setup was complemented with a 2155 column oven by PHARMACIA LKB (Upsala, Sweden). If not otherwise noted, the purification was conducted at room temperature.

The following columns were used for compound retention:

Column 1:	ACE Excel 2 C18	100A-2-C18 100 mm x 2.1 mm, 2 µm	UPLC
Column 2:	MN Nucleodur RP-C18-ec	100-5-C18, 250 mm x 4.6 mm, 5 µm	analytical
Column 3:	MN Nucleodur RP-C18-ec	100-5-C18, 250 mm x 10 mm, 5 µm	semi-prep.
Column 4:	MN Nucleodur RP-C18-ec	300-5-C18, 250 mm x 10 mm, 5 µm	semi-prep.
Column 5:	MN Nucleodur RP-C18-ec	100-5-C18, 250 mm x 21 mm, 5 µm	preparative
Column 6:	ACE Excel 2 C18	100A-5-C18, 150 mm x 21.2 mm, 5µm	preparative

Applying a flow rate of 0.30-0.8 mL/min (UPLC), 1 mL/min (analytical), 3 mL/min (semipreparative) or 10 mL/min (preparative) a linear gradient of solvent A to solvent B from the eluent systems listed below was conducted.

Eluent system	Solvent A	Solvent B
I	H₂O + 0.1% TFA	MeCN + 0.085% TFA
II	H₂O + 0.1% TFA	MeOH + 0.085% TFA
III	H₂O + 0.1% TFA	MeCN/H₂O 4:1 + 0.1% TFA
IV	H₂O + 0.1% TFA	MeCN + 0.1% TFA
V	H₂O + 0.1% TFA	MeOH + 0.1% TFA
VI	H₂O + 0.1% TFA	MeOH/1-PrOH 4:1 + 0. 1% TFA
VII	TEAA buffer 0.1 M pH 7.0	MeCN/TEAA buffer 3:2
VIII	H₂O + 0.1% TFA	MeOH/1-PrOH 3:2 + 0.1% TFA

Peptides constructed only from proteinogenic amino acids were detected at 215 nm, 254 nm, and 280 nm. NVOC and DMNPB containing compounds were observed at 347 nm while chromatograms of coumarin derivatives were recorded at 390 nm. UPLC analysis was conducted at 50 °C.

Prior to injection the analytes were dissolved in a minimal amount of a solvent mixture corresponding to the starting conditions of the HPLC run. Poorly soluble transmembrane peptides were dissolved in HFIP and then diluted to the fivefold volume with H₂O to avoid premature elution. The solutions were filtrated through a *Chromafil* syringe filter (0.45µm pore size) by *Macherey Nagel* (Düren, Germany).

7.3.4 LC-MS

LC-MS measurements were conducted on an LTQ XL ion trap mass spectrometer from *Thermo Fischer Scientific* (Waltham, USA) connected to an *UltiMate 3000* UPCL system from *Thermo Fischer Scientific* (Waltham, USA) consisting of autosampler-, pumps-, column oven-, detector, and diode array modules of the *3000* series. As eluents, H_2O (LC-MS grade) + 0.1% FA (solvent A) and MeCN (LC-MS grade) + 0.1% FA (solvent B) were used. Separation was achieved by an ACE Excel (100A-2-C18 100 mm x 2.1 mm, 2 µm) column and a linear gradient of solvents A and B using a flow rate of 0.35 mL/min at 40 °C. UV detection was performed at the wavelengths described above. Chromatograms are mostly shown as a representation of the total ion count (TIC).

7.4 Fusion experiments

7.4.1 Preparation of peptide/lipid films

Stock solutions of unlabeled lipids 1,2-dioleoyl-sn-glycero-3-phosphocholine (18:1, Δ9-cis, DOPC), 1,2-dioleoylsn-glycero-3-phosphoethanolamine (18:1, Δ9-cis, DOPE) and cholesterol were prepared in chloroform at a concentration of 20 mg/mL. Labeled lipids 1,2-dioleoyl-sn-glycero-3-phosphoethanolamine-N-(7-nitro-2-1,3-benzoxadiazol-4-yl) (ammonium salt) (NBD-DOPE) and 1,2-dioleoyl-sn-glycero-3-phosphoethanolamine-N-(lissamine rhodamine B sulfonyl) (ammonium salt) (Rh-DOPE) were prepared as 2 mg/mL stock solutions and stored in brown glass vials. Peptides were dissolved in TFE and their concentration was measured by UV/VIS at 280 nm for uncaged peptides and at the absorption maximum of the corresponding PPG for caged peptides. Peptide/lipid films were prepared in small glass test tubes and during mixing all test tubes, stock solutions and solvents were cooled with ice. Appropriate volumes of lipid solutions were combined to a total lipid amount of 2.5 µmol. For unlabeled peptide/lipid films the lipids were mixed to give a ratio of DOPC/DOPE/Cholesterol 50:25:25. Labeled peptide/lipid mixtures were combined at a ratio of DOPC/DOPE/Cholesterol/NBD-DOPE/Rh-DOPE 50:22:25:1.5:1.5. The lipid mixtures were topped with cholesterol to give a total volume of 500 µL. Peptides were added to the lipid solution to give a lipid/ peptide ratio of 200/1 and the solutions were topped with TFE to a total TFE volume of 500 µL. The mixtures were allowed to warm up to room temperature and vortexed for 10 s. Then, the solutions were heated to 50 °C and the solvents were removed in a faint N_2 stream. The resulting clear films on the bottom of the test tubes were stored in a vacuum oven at 50 °C for an least 12 h before use to complete solvent removal.

7.4.2 Vesicle preparation by extrusion

Vesicles were prepared by extrusion of rehydrated peptide/lipid films[231] through a polycarbonate membrane (100 nm pore diameter, 19 mm disc diameter) by *Avestin* (Ottawa, Canada) with Whatman polyester drain discs by GE healthcare (Little Chalfont, UK) mounted on both sides of the membrane in a *LiposoFast-Basic* extruder by *Avestin*. Before the extruder was assembled and equilibrated by passing through HEPES buffer (20 mM, 100 mM KCl, 1 mM EDTA, 1 mM DTT, pH 7.4). The buffer was prepared freshy on the day of use by diluting a 10-fold stock solution of HEPES, KCl and EDTA with ultrapure water and adjusting the pH with 1 M KOH. After addition of DTT the buffer was thoroughly mixed and passed through a *Chromafil syringe* filter (0.45µm pore size) by *Macherey Nagel* (Düren, Germany). 500 µL HEPES buffer and ~10 glass beads

(Ø 2mm) were added to the peptide/lipid films. The films were incubated at 50 °C for 1 h and subsequently resuspended by vortexing for 30 s and treatment in an ultrasonic bath for few seconds. The homogenous emulsion was loaded into one syringe of the prepared extruder and extruded 31 times. Between different lipid films, the extruder was completely disassembled and rinsed with EtOH and ultrapure water.

7.4.3 Total lipid mixing assay

Detection of fluorescence change upon lipid mixing was carried out on a JASCO spectrofluorometer. Donor fluorophores were excited at 460 nm and emission was monitored at 535 nm in a time course measurement over 20 min to 2 h. At the beginning of each experiment, 1300 µL of HEPES buffer were added to a stirrable fluorescence cuvette (*Hellma analytics*, semi-micro, 1500 µL) and 40 µL of labeled vesicle suspension were added and an emission spectrum (ext. 460 nm) was measured to check the liposome quality. The time course measurement was started and after 180 s the unlabeled vesicles were added in 4-fold excess (160 µL) to obtain a theoretical final lipid concentration of 180 µM. For photocleavage of PPGs by method a), the cuvette was removed from the spectrometer and carried to the uncaging setup in an adjacent laboratory. By method c) the uncaging was performed directly in the spectrometer after opening the lid. In the end of the experiment, 25 µL TX-100 (10% in HEPES buffer (*v/v*)) were added and the fluorescence at 535 nm was observed until equilibration of the signal.

The fluorescence signal was normalized by the following equation:

$$F\ [\%] = \frac{F_t - F_0}{F_{total} - F_o} \cdot 100$$

Where F_t is the measured signal, F_0 is calculated from the average signal of 30 s before addition of unlabeled vesicles and F_{total} is calculated from the average 30 s of fluorescence signal after addition of TX-100. The timeline was adjusted to start with $t = 0$ s at the first measured value after addition of the unlabeled vesicles.

3-(3,4-dimethoxyphenyl)butan-2-one

$C_{11}H_{14}O_3$
194.2 g/mol
9

$C_{12}H_{16}O_3$
208.3 g/mol
10

Under inert atmosphere 3,4-dimethoxyphenylacetone (10.0 g, 55.0 mmol, 1.0 eq) was added dropwise to a suspension of sodium hydride (2.10 g NaH 60% in paraffin, 52.0 mmol, 1.0 eq) in dry THF (75 mL) and stirred for 30 min at room temperature. After cooling with an ice bath to 0 °C methyl iodide was added, and the suspension was stirred for 30 min at 0 °C and for 1 h at rt. The reaction mixture was quenched by addition of sat. aq. NaHCO$_3$ (100 mL) and extracted with EtOAc (2 x 200 mL). The combined organic layers were washed with brine (2 x 200 mL) and dried over MgSO$_4$. After removal of the solvent under reduced pressure the product (8.70 g, 42.0 mmol, 81%) was obtained as a yellow oil and used without further purification.

TLC: R_f = 0.19 (pentane/EtOAc 6:1).

¹H-NMR (400 MHz, CDCl$_3$): δ (ppm) = 6.83 (d, $^3J_{HH}$ = 8.2 Hz, 1 H, CH_{ar}), 6.77 (dd, ^3J = 8.1 Hz, ^4J = 2.0 Hz, 1 H, CH_{ar}), 6.68 (d, $^4J_{HH}$ = 2.1 Hz, 1 H, CH_{ar}), 3.86 (s, 6 H, 2 x OCH_3), 3.68 (q, ^3J = 7.1 Hz, 1 H, CH-3), 2.05 (s, 3 H, CH_3-1), 1.37 (d, ^3J = 7.1 Hz, 3 H, CH_3-4).

¹³C-NMR (100 MHz, CDCl$_3$): δ (ppm) = 209.0 (CO), 149.3 ($C_{q,ar}$), 148.4 ($C_{q,ar}$), 132.9 ($C_{q,ar}$), 120.0 (CH_{ar}), 111.6 (CH_{ar}), 110.6 (CH_{ar}), 56.0 (2 x OCH_3), 53.2 (CH-3), 28.4 (CH_3-1), 17.1 (CH_3-4).

ESI-MS m/z : 231.1 [M+Na]$^+$, 439.2 [2M+Na]$^+$.

HR-MS (ESI): calcd. for $C_{12}H_{16}NaO_3$, ([M+Na]$^+$): 231.0996, found: 231.0992.

3-(4,5-dimethoxy-2-nitrophenyl)butan-2-one

$C_{12}H_{16}O_3$
208.3 g/mol
10

$C_{12}H_{15}NO_5$
253.3 g/mol
11

To an ice-cold solution of **10** (5.00 g, 24.0 mmol, 1.0 eq) in acetic acid (50 mL) fuming nitric acid (4.8 mL, 0.12 mol, 4.8 eq) was added dropwise. After stirring the solution for 10 min the ice bath was removed and stirring was continued for 20 min. The reaction mixture was poured over crushed ice (200 mL) and after complete melting of the ice extracted with ethyl acetate (3 x 100 mL). The combined organic layers were washed with sat. aq. $NaHCO_3$ (2 x 100 mL) and brine (2 x 100 mL) and dried over $MgSO_4$. Removal of the solvent yielded the product (5.90, 23.3 mmol, 97%) as a brown solid.

TLC: R_f = 0.23 (pentane/EtOAc 5:1).

^{1}H-NMR (300 MHz, $CDCl_3$): δ (ppm) = 7.55 (s, 1 H, CH_{ar}), 6.68 (s, 1 H, CH_{ar}), 4.51 (q, ^3J = 7.1 Hz, 1 H, CH-3), 3.91 (s, 6 H, 2 x OCH_3), 2.17 (s, 3 H, CH_3-1), 1.44 (d, ^3J = 7.1 Hz, 3 H, CH_3-4).

^{13}C-NMR (125 MHz, $CDCl_3$): δ (ppm) = 207.4 (CO), 153.1 ($C_{q,ar}$), 147.6.4 ($C_{q,ar}$), 141.1 ($C_{q,ar}$), 123.0 ($C_{q,ar}$), 110.5 ($C H_{ar}$), 108.2 ($C H_{ar}$), 56.3 (2 x O$C H_3$), 48.1 (CH-3), 28.8 ($C H_3$-1), 16.8 ($C H_3$-4).

ESI-MS m/z: 276.1 [M+Na]$^+$, 529.2 [2M+Na]$^+$.

HR-MS (ESI): calcd. for $C_{12}H_{16}NO_5$, ([M+H]$^+$): 254.1023, found: 254.1019.

3-(4,5-dimethoxy-2-nitrophenyl)butan-2-ol

O
NO$_2$
C$_{12}$H$_{15}$NO$_5$
253.3 g/mol
11

OH
NO$_2$
C$_{12}$H$_{17}$NO$_5$
255.3 g/mol
12

Sodium borohydride (2.1 g, 55.0 mmol, 1.4 eq) was added to a solution of **11** (10.0 g, 39.0 mmol, 1.0 eq) in THF/2-propanol (3:2, 50 mL). After 1 h of stirring the reaction was quenched with aq. HCl (1 M, 100 mL) and extracted with EtOAc (3 x 100 mL). The combined organic layers were washed with sat. NaHCO$_3$ solution (100 mL) and brine (100 mL). Removal of the solvent yielded the product (7.00 g, 28.0 mmol, 72%) as a mixture of diastereomers.

TLC: R_f = 0.20 (pentane/EtOAc 5:1).

^{1}H-NMR (300 MHz, CDCl$_3$): δ (ppm) = 7.41-7.33 (m, 1 H, CH_{ar}), 6.97-6.88 (m, 1 H, CH_{ar}), 3.99-3.85 (m, 7 H, 2 x OCH3, CH-2), 3.58-3.45 (m, 1 H, CH-3), 1.34-1.12 (m, 6 H, CH_3-4, CH_3-1).

^{13}C-NMR (100 MHz, CDCl$_3$): δ (ppm) = 152.6 ($C_{q,ar}$), 147.0 ($C_{q,ar}$), 141.1 ($C_{q,ar}$), 123.0 ($C_{q,ar}$), 110.5 (CH_{ar}), 108.2 (CH_{ar}), 72.3 (CH-2), 56.3 (2 x OCH_3), 48.1 (CH-2), 28.8 (CH_3-1), 16.8 (CH_3-4).

ESI-MS m/z : 278.1 [M+Na]$^+$, 533.2 [2M+Na]$^+$.

HR-MS (ESI): calcd. for C$_{12}$H$_{17}$N$_1$NaO$_5$, ([M+H]$^+$): 278.0999, found: 278.1001.

Boc-L-Glu(DMNPB)-OtBu

3-(4,5-dimethoxy-2-nitrophenyl)butan-2-ol (8.00 g, 31.3 mmol, 1 eq) was dissolved in dry DCM (300 mL) under inert atmosphere. Boc-L-Glu-OtBu (10.4 g, 34.4 mmol, 1.1 eq), DCC (7.1 g, 34.4 mmol, 1.1 eq) and DMAP (0.31 g, 2.5 mmol, 0.08 eq) were added. The reaction mixture was stirred at room temperature for 12 h, quenched with sat. NaHCO$_3$ (200 mL) and extracted with EtOAc (3 x 150 mL). The combined organic layers were washed with water (200 mL) and brine (200 mL) and dried over MgSO$_4$. After removal of the solvent under reduced pressure the crude product was purified by column chromatography (pentane/EtOAc 4:1). The solid product (14.0 g, 26 mmol, 83%) was obtained as a mixture of diastereomers.

TLC: R_f = 0.17 (pentane/EtOAc 4:1).

^{1}H-NMR (300 MHz, CDCl$_3$): δ (ppm) = 7.39-7.33 (m, 1 H, CH_{ar}), 6.85-6.75 (m, 1 H, CH_{ar}), 5.17-5.08 (m, 1 H, CH-2$_{DMNPB}$), 3.97-3.87 (m, 6 H, 2 x OCH_3), 3.78-3.69 (m, 1 H, CH-3$_{DMNPB}$), 2.24-1.86 (m, 4 H, δ-CH$_2$, γ-CH$_2$), 1.46-1.38 (m, 18 H, CH$_{3,Boc}$, CH$_{3,tBu}$), 1.34-1.28 (m, 3 H, CH_3-1$_{DMNPB}$), 1.25-1.21 (m, 3 H, CH_3-4$_{DMNPB}$).

^{13}C-NMR (100 MHz, CDCl$_3$): δ (ppm) = 172.2 (γ-CO), 172.2 (α-CO), 155.4 ($C_{q,arom}$), 152.6 (CO$_{Boc\text{-}carbamat}$), 147.2 ($C_{q,arom}$), 142.8 ($C_{q,ar}$), 132.3 (CH_{ar}), 121.5 ($C_{q,ar}$), 112.0 (CH_{ar}), 82.30 (C$H_{3,tBu}$), 79.7 (C$H_{3,Boc}$), 73.9 (CH-2$_{DMNPB}$), 56.5 (α-CH), 56.4 (OCH_3), 56.3 (OCH_3), 38.5 (CH-3$_{DMNPB}$), 30.7 (γ-CH$_2$), 28.41 (CH$_{3,tBu}$), 28.11 (CH$_{3,Boc}$), 21.2 (β-CH$_2$), 18.6 (CH$_3$-4$_{DMNPB}$), 16.97 (CH$_3$-4$_{DMNPB}$).

ESI-MS m/z : 563.3 [M+Na]$^+$, 541.3 [M+H]$^+$.

HR-MS (ESI): calcd. for C$_{26}$H$_{40}$N$_2$NaO$_{10}$, ([M+Na]$^+$): 563.2575, found: 563.2578.

H-L-Glu(DMNPB)-OH

O₂N

C₂₆H₄₀N₂O₁₀
540.6 g/mol
13

O₂N
H₂N
OH

C₁₇H₂₄N₂O₈
384.4 g/mol
90

13 (14.0 g, 26 mmol, 1 eq) was dissolved TFA/DCM (1:1, 120 mL) and stirred at room temperature for 1 h. The solvents were removed in an N_2 stream and residual TFA was coevaporated with DCM (20 mL). The product (9.90 g, 26.0 mmol, quant.) was obtained as a light brown solid.

¹H-NMR (300 MHz, CDCl₃): δ (ppm) = 7.43-7.21 (m, 1 H, C*H*ar), 6.83-6.63 (m, 1 H, C*H*ar), 5.14-5.04 (m, 1 H, C*H*-2DMNPB), 3.94-3.86 (m, 6 H, 2 x OC*H*₃), 3.78-3.69 (m, 1 H, C*H*-3DMNPB), 2.65-2.10 (m, 4 H, δ-CH₂, γ-CH₂), 1.32-1.28 (C*H*₃-1DMNPB), 1.21-1.04 (m, 3 H, C*H*₃-4DMNPB).

¹³C-NMR (100 MHz, CDCl₃): δ (ppm) = 174.0 (*C*O₂H), 173.5 (γ-*C*O), 152.8 (*C*q,arom), 147.5 (*C*q,arom), 147.4 (*C*q,arom), 131.9 (*C*Har), 118.7 (*C*q,arom), 114.0 (*C*Har), 75.8 (*C*H-2DMNPB), 56.4 (α-*C*H), 56.3 (2 x O*C*H₃), 37.9 (*C*H-3DMNPB), 27.8 (γ-*C*H₂), 24.8 (β-*C*H₂), 17.3 (*C*H₃-4DMNPB), 16.6 (*C*H₃-4DMNPB).

ESI-MS *m/z* = 385.2 [M+H]⁺, 407.2 [M+Na]⁺.

HR-MS (ESI): calcd. for C₁₇H₂₃N₂O₈, ([M+H]⁺): 385.1605, found: 385.1605.

Fmoc-L-Glu(DMNPB)-OH

C$_{17}$H$_{24}$N$_2$O$_8$
384.4 g/mol
90

C$_{32}$H$_{34}$N$_2$O$_{10}$
606.6 g/mol
8

90 (9.90 g, 26 mmol, 1 eq) was dissolved in aqueous K$_2$CO$_3$ solution (9%, 100 mL). Fmoc-OSu (8.77 g, 26 mmol, 1 eq) in dioxane (100 mL) was added at 0 °C. The reaction mixture was stirred for 1 h at 0 °C and allowed to come to room temperature overnight. The mixture was extracted with diethyl ether (3 x 100 mL) and the combined organic layers were discarded. The aqueous phase was adjusted to pH 2 with aqueous HCl (2 M) and extracted with EtOAc (3 x 100 mL). The combined organic layers were washed with water (100 mL) and brine (100 mL) and dried over MgSO$_4$. After removal of the solvent under reduced pressure the crude product was purified by column chromatography (pentane/EtOAc/AcOH, 70:30:1). The product (6.80 g, 11.0 mmol, 42%) was obtained as a light brown solid.

TLC: R_f = 0.13 (pentane/EtOAc/AcOH, 70:30:1).

^{1}H-NMR (300 MHz, CDCl$_3$): δ (ppm) = 7.73 (d, ^3J = 7.4 Hz, 2 H, 2 x C$H_{ar,Fmoc}$), 7.60-7.53 (m, 2 H, 2 x C$H_{ar,Fmoc}$), 7.40-7.25 (m, 5 H, 4 x C$H_{ar,Fmoc}$, C$H_{ar,DMNPB}$), 6.83-6.73 (m, 1 H, C$H_{ar,DMNPB}$), 5.16-5.03 (m, 1 H, CH-2$_{DMNPB}$), 4.44-4.32 (m, 2 H, C$H_{2,sp3,Fmoc}$), 4.28-4.07 (m, 2 H, C$H_{sp3,Fmoc}$, α-CH), 3.94-3.79 (m, 6 H, 2 x OCH_3), 3.78-3.69 (m, 1 H, CH-3$_{DMNPB}$), 2.47-1.64 (m, 4 H, δ-CH_2, γ-CH_2), 1.35-1.00 (m, 6 H, CH_3-1$_{DMNPB}$, CH_3-4$_{DMNPB}$).

^{13}C-NMR (125 MHz, CDCl$_3$): δ (ppm) = 176.0 (CO$_2$H), 172.5 (γ-CO), 162.8 (CO$_{Fmoc-carbamate}$), 156.0 ($C_{q,arom}$), 153.5 ($C_{q,ar}$), 147.2 ($C_{q,ar}$), 143.9 ($C_{q,ar}$), 141.4 ($C_{q,ar}$), 132.1 (CH$_{ar}$), 127.8 (CH$_{ar}$), 127.2 (CH$_{ar}$), 125.2 (CH$_{ar}$), 118.7 ($C_{q,ar}$), 109.8 (CH$_{ar}$), 107.7 ($C_{q,ar}$), 74.4 (CH-2$_{DMNPB}$), 67.3 (CH$_{2,Fmoc}$), 56.4 (α-CH), 56.2 (2 x OCH$_3$), 47.1 (CH$_{sp3,Fmoc}$), 38.1 (CH-3$_{DMNPB}$), 30.6 (γ-CH$_2$), 27.4 (β-CH$_2$), 18.3 (CH$_3$-4$_{DMNPB}$), 17.8 (CH$_3$-4$_{DMNPB}$).

ESI-MS m/z = 629.2 [M+Na]$^+$, 1235.4 [2M+Na]$^+$.

HR-MS (ESI): calcd. for C$_{32}$H$_{34}$N$_2$NaO$_{10}$, ([M+Na]$^+$): 629.2106, found: 629.2098.

7-Diethylamino-4-hydroxymethylcoumarin

$C_{14}H_{17}NO_2$
231.3 g/mol
15

$C_{14}H_{17}NO_3$
247.3 g/mol
16

7-Diethylamino-4-methylcoumarin (4.21 g, 18.2 mmol, 1.0 eq) and selenium dioxide (2.10 g, 18.9 mmol, 1.04 eq) were dissolved in xylene (350 mL) and heated to reflux for 48 h. The completely cooled mixture was filtered, and the filtrate was reduced *in vacuo*. The residue was dissolved in methanol (400 mL), NaBH₄ (0.64 g, 17.0 mmol, 0.9 eq) was added and the mixture was stirred at room temperature for 4 h. Water (150 mL) was added and the unreacted NaBH₄ was quenched by neutralizing with 1 M HCl. Methanol was removed at reduced pressure and the aqueous phase was extracted with DCM (5 x 100 mL). The combined organic layers were dried over MgSO₄ and the solvent was removed under reduced pressure. The crude product was purified by column chromatography (gradient of DCM to DCM/acetone 1:1). The product (1.92 g, 7.8 mmol, 43%) was obtained as a brown solid.

TLC: R_f = 0.35 (DCM/acetone 10:1)

^{1}H-NMR (300 MHz, CDCl₃): δ (ppm) = 7.31 (d, $^3J_{HH}$ = 9.0 Hz, 1 H, H-5$_{coum}$), 6.55 (dd, $^3J_{HH}$ = 9.0 Hz, $^4J_{HH}$ = 2.6 Hz, 1 H, H-6$_{coum}$), 6.48 (d, $^4J_{HH}$ = 2.6 Hz, 1 H, H-8$_{coum}$), 6.26 (t, $^4J_{HH}$ = 1.3 Hz, 1 H, H-3$_{coum}$), 4.82 (d, J = 4.4 Hz, 1 H, CH_2OH), 3.44 (q, J = 7.1 Hz, 4 H, CH_2CH₃), 2.61 (t, J = 5.6 Hz, 1 H, OH), 1.19 (t, J = 7.1 Hz, 6 H, CH₂CH_3).

^{13}C-NMR (75 MHz, CDCl₃): δ (ppm) = 162.7 (C-2$_{coum}$), 156.1 (C-8a$_{coum}$), 154.9 (C-7$_{coum}$), 150.5 (C-4$_{coum}$), 124.4 (C-5$_{coum}$), 108.6 (C-3$_{coum}$), 106.3 (C-4a$_{coum}$), 105.3 (C-6$_{coum}$), 97.7 (C-8$_{coum}$), 60.9 (CH₂OH), 44.7 (CH₂CH₃), 12.40 (CH₂CH₃).

ESI-MS *m/z*: 248.1 [M+H]⁺, 517.2 [2M+Na]⁺, 764.4 [3M+Na]⁺.

HR-MS (ESI): calcd. for $C_{14}H_{17}NO_3$, ([M+H]⁺): 248.1281, found: 248.1283.

(7-Diethylaminocoumarin-4-yl)methyl (4-nitrophenyl) carbonate

$C_{14}H_{17}NO_3$
247.3 g/mol
16

$C_{21}H_{20}N_2O_7$
412.4 g/mol
17

16 (0.99 g, 4.0 mmol, 1.0 eq) and 4-nitrophenyl chloroformate (1.25 g, 6.2 mmol, 1.5 eq) were suspended in DCM (25 mL), cooled with an ice bath and DIPEA (0.70 mL, 4.0 mmol, 1.0 eq) was added yielding a clear solution. The reaction mixture was stirred for 8 h, allowing it to come to room temperature gradually. The newly formed precipitate was dissolved by addition of DIPEA (0.70 mL, 4.0 mmol, 1.0 eq) and the solution was stirred for 1 h. The reaction mixture was directly adsorbed on silica and the crude product was purified by column chromatography (gradient DCM to DCM/acetone 10:1). The product (0.85 g, 2.0 mmol, 50%) was obtained as a yellow solid.

TCL: R_f = 0.14 (DCM)

^{1}H-NMR (300 MHz, CDCl$_3$): δ (ppm) = 8.30 (dt, $^3J_{HH}$ = 9.2 Hz, $^4J_{HH}$ = 2.2 Hz, 2 H, H-3$_{pNP}$), 7.42 (dt, $^3J_{HH}$ = 9.2 Hz, $^5J_{HH}$ = 2.2 Hz, 1 H, H-2$_{pNP}$), 7.32 (d, $^3J_{HH}$ = 9.0 Hz, 1 H, H-5$_{coum}$), 6.62 (dd, $^3J_{HH}$ = 9.0 Hz, $^4J_{HH}$ = 2.6 Hz, 1 H, H-6$_{coum}$), 6.53 (d, $^4J_{HH}$ = 2.6 Hz, 1 H, H-8$_{coum}$), 6.23 (t, $^4J_{HH}$ = 1.1 Hz, 1 H, H-3$_{coum}$), 5.41 (d, $^4J_{HH}$ = 1.1 Hz, 1 H, CH_2O), 3.42 (q, $^3J_{HH}$ = 7.1 Hz, 4 H, CH_2CH$_3$), 1.22 (t, $^3J_{HH}$ = 7.1 Hz, 6 H, CH$_2$CH_3).

^{13}C-NMR (75 MHz, CDCl$_3$): δ (ppm) = 161.9 (C-2$_{coum}$), 156.2 (C-1$_{pNP}$), 155.1 (C-8a$_{coum}$), 152.0 (CO$_{carbonate}$), 150.8 (C-7$_{coum}$), 148.0 (C-4$_{coum}$), 145.6 (C-4$_{pNP}$), 126.1 (C-5$_{coum}$), 125.3 (C-3$_{pNP}$), 121.6 (C-2$_{pNP}$), 115.6 (C-3$_{coum}$), 109.0 (C-4a$_{coum}$), 106.7 (C-6$_{coum}$), 97.9 (C-8$_{coum}$), 65.7 (CH$_2$O), 44.9 (CH$_2$CH$_3$), 12.5 (CH$_2$CH$_3$).

ESI-MS m/z : 413.1 [M+H]$^+$, 435.1 [M+Na]$^+$.

HR-MS (ESI): calcd. for $C_{21}H_{21}N_2O_7$, ([M+H]$^+$): 413.1346, found: 248.1343; calcd. for $C_{21}H_{20}N_2NaO_7$, ([M+Na]$^+$): 435.1163, found: 435.1163.

Fmoc-L-Lys(DEACM)-OH

17 (0.850 g, 2.07 mmol, 1.0 eq) suspended in DMF/DCM (1:1, 30 mL) was combined with Fmoc-L-Lys-OH (0.737 g, 2.00 mmol, 0.97 eq) suspended in toluene/DCM (3:2, 65 mL) and cooled in an ice bath. DIPEA (347 µL, 2.0 mmol, 1.0 eq) was added and the reaction mixture was stirred for 12 h allowing it to slowly come to room temperature. Further DIPEA (100 µL, 0.6 mmol, 0.3 eq) was added and the reaction mixture was stirred at rt for 1 h. The solvents were removed under reduced pressure. The residue was dissolved in DCM (100 mL), adsorbed to silica (5 g), and purified by column chromatography (DCM/MeOH/AcOH, 95:5:0.1). After lyophilization from dioxane the product (1.12 g, 1.70 mmol, 87%) was obtained as a yellow powder.

TLC: R_f = 0.26 (DCM/MeOH/AcOH, 95:5:0.1)

^{1}H-NMR (300 MHz, DMSO-d$_6$): δ (ppm) = 7.86 (d, $^3J_{HH}$ = 7.4 Hz, 2 H, C$H_{ar,Fmoc}$), 7.71 (d, $^3J_{HH}$ = 7.4 Hz, 2 H, C$H_{ar,Fmoc}$), 7.43-7.36 (m, 3 H, H-5$_{coum}$, C$H_{ar,Fmoc}$), 7.32 (dt, $^3J_{HH}$ = 7.4 Hz, $^4J_{HH}$ = 1.1 Hz, 2 H, C$H_{ar,Fmoc}$), 6.65 (dd, $^3J_{HH}$ = 9.0 Hz, $^4J_{HH}$ = 2.5 Hz, 1 H, H-6$_{coum}$), 6.51 (d, $^4J_{HH}$ = 2.5 Hz, 1 H, H-8$_{coum}$), 5.98 (s, 1 H, H-3$_{coum}$), 5.19 (s, 1 H, CH_2O), 4.32 (m, 3 H, C$H_{sp3,Fmoc}$, C$H_{2,Fmoc}$), 4.00-3.91 (m, 2 H, α-CH_2), 3.40 (q, $^3J_{HH}$ = 7.0 Hz, 4 H, CH_2CH$_3$), 3.09-3.00 (m, 2 H, ε-CH_2), 1.81-1.55 (m, 2 H, β-CH_2), 1.53-1.23 (m, 4 H, γ-CH_2, δ-CH_2), 1.22 (t, $^3J_{HH}$ = 7.1 Hz, 6 H, CH$_2$CH_3).

^{13}C-NMR (125 MHz, CDCl$_3$): δ (ppm) = 173.6 (CO$_2$H), 161.5 (C-2$_{coum}$), 155.9 (CO$_{coum-carbamate}$) 155.5 (C-8a$_{coum}$), 155.2 (CO$_{Fmoc-carbamate}$), 151.6 (C-7$_{coum}$), 150.2 (C-4$_{coum}$), 143.6 ($C_{q,Fmoc}$), 140.5 (Fmoc-$C_{q,Fmoc}$), 127.4 (CH$_{ar,Fmoc}$), 126.8 (CH$_{ar,Fmoc}$), 125.0 (CH$_{ar,Fmoc}$, C-5$_{coum}$), 119.8 (CH$_{ar,Fmoc}$), 108.6 (C-3$_{coum}$), 105.2 (C-6$_{coum}$), 104.4 (C-4a$_{coum}$), 96.7 (C-8$_{coum}$), 65.5 (CH$_{2,Fmoc}$), 60.7 (CH$_2$O), 53.7 (α-CH$_2$), 46.6 (CH$_{sp3,Fmoc}$), 43.9 (CH$_2$CH$_3$), 40.1 (ε-CH$_2$), 30.4 (β-CH$_2$), 28.8 (δ-CH$_2$), 22.8 (γ-CH$_2$), 12.2 (CH$_2$$CH_3$).

ESI-MS m/z: 413.1 [M+H]$^+$, 435.1 [M+Na]$^+$.

HR-MS (ESI): calcd. for C$_{36}$H$_{40}$N$_3$O$_8$, ([M+H]$^+$): 642.2816, found: 642.2810; calcd. for C$_{36}$H$_{39}$N$_3$NaO$_8$, ([M+Na]$^+$): 664.2642, found: 664.2629.

Fmoc-L-Glu(DEACM)-O^{t}Bu

$C_{14}H_{17}NO_3$
247.3 g/mol
16

$C_{38}H_{42}N_2O_8$
654.8 g/mol
22

To a solution of **16** (0.60 g, 2.09 mmol, 1.00 eq) and Fmoc-L-Glu-O^{t}Bu (0.98 g, 2.30 mmol, 1.10 eq) in anhydrous DCM (50 mL) DCC (0.47 g, 2.30 mmol, 1.10 eq) and DMAP (26 mg, 0.21 mmol, 0.10 eq) were added. The solution was stirred for 12 h at room temperature. Then, the reaction mixture was washed with saturated NaHCO$_3$ solution (3 x 50 mL), dried over MgSO4, and adsorbed onto silica (3 g). Purification by column chromatography (DCM/MeOH, 99:1) yielded the product (1.03 g, 1.93 mmol, 96%) as a yellow solid.

TLC: R_f = 0.55 (DCM/MeOH, 98:2).

^{1}H-NMR (300 MHz, DMSO-d6): δ (ppm) = 7.87 (d, $^3J_{HH}$ = 7.5 Hz, 2 H, C$H_{ar,Fmoc}$), 7.71 (d, $^3J_{HH}$ = 7.4 Hz, 2 H, C$H_{ar,Fmoc}$), 7.66 (d, $^3J_{HH}$ = 8.04 Hz, 1 H, NH), 7.44 (d, $^3J_{HH}$ = 9.03 Hz, 1 H, H-5$_{coum}$), 7.40 (t, $^3J_{HH}$ = 7.4 Hz, 2 H, C$H_{ar,Fmoc}$), 7.31 (dt, $^3J_{HH}$ = 7.4 Hz, $^4J_{HH}$ = 0.7 Hz, 2 H, C$H_{ar,Fmoc}$), 6.66 (dd, $^3J_{HH}$ = 9.0 Hz, $^4J_{HH}$ = 2.6 Hz, 1 H, H-6$_{coum}$), 6.53 (d, $^3J_{HH}$ = 2.5 Hz, 1 H, H-8$_{coum}$), 6.00 (s, 1 H, H-3$_{coum}$), 5.28 (s, 2 H, CH_2O), 4.35-4.17 (m, 3 H, α-CH, C$H_{2,Fmoc}$), 4.04-3.94 (m, 1 H, C$H_{sp3,Fmoc}$), 3.41 (q, $^3J_{HH}$ = 7.0 Hz, 4 H, CH_2CH$_3$), 2.60-2.50 (m, 2 H, γ-CH_2), 1.98-1.64 (m, 2 H, β-CH_2), 1.47 (s, 3 H, ^{t}Bu-CH_3) 1.19 (t, $^3J_{HH}$ = 7.2 Hz, 6 H, CH$_2$CH_3).

^{13}C-NMR (125 MHz, DMSO-d6): δ (ppm) = 171.4 (CO$_2$H), 170.7 (γ-CO), 160.3 (C-2$_{coum}$), 155.8 (CO$_{Fmoc-carbamate}$), 155.5 (C-8a$_{coum}$), 150.2 (C-4$_{coum}$), 150.1 (C-7$_{coum}$), 143.5 ($C_{q,Fmoc}$), 140.5 ($C_{q,Fmoc}$), 127.3 (CH$_{ar,Fmoc}$), 126.8 (CH$_{ar,Fmoc}$), 125.2 (C-5$_{coum}$), 124.9 (CH$_{ar,Fmoc}$), 119.8 (CH$_{ar,Fmoc}$), 108.5 (C-6$_{coum}$), 105.1 (C-4a$_{coum}$, C-3$_{coum}$), 96.7 (C-8$_{coum}$), 80.6 (^{t}Bu-C$_q$), 65.5 (C$H_{2,Fmoc}$), 61.1 (CH_2O$_{coum}$), 53.5 (α-CH), 46.6 (C$H_{sp3,Fmoc}$), 43.8 (CH_2CH$_3$), 33.2 (γ-CH$_2$), 27.5 (^{t}Bu-CH$_3$), 24.3 (β-CH$_2$), 12.2 (CH$_2$$CH_3$).

ESI-MS m/z : 695.3 [M+H]$^+$, 717.3 [M+Na]$^+$.

HR-MS (ESI): calcd. for $C_{41}H_{47}N_2O_8$, ([M+H]$^+$): 695.3327, found: 639.3325; calcd. for $C_{41}H_{46}N_2NaO_8$, ([M+Na]$^+$): 717.3136, found: 717.3136.

Fmoc-L-Glu(DEACM)-OH

C$_{38}$H$_{42}$N$_2$O$_8$
654.8 g/mol
22

C$_{34}$H$_{34}$N$_2$O$_8$
598.7 g/mol
23

22 (1.17 g, 1.79 mmol, 1.00 eq) was dissolved in DCM/TFA (1:1, 60 mL) and stirred at rt for 1 h. The solvents were removed in a N$_2$ stream, residual TFA was coevaporated under reduced pressure with DCM (3 x 50 mL). The product (1.07 g, 1.79 mmol, quant.) was obtained as a yellow solid.

TLC: R_f = 0.32 (DCM/MeOH/AcOH 95:5:0.1).

^{1}H-NMR (400 MHz, CD$_3$OD): δ (ppm) = 7.71 (d, $^3J_{HH}$ = 7.5 Hz, 2 H, C$H_{ar,Fmoc}$), 7.60 (dd, $^3J_{HH}$ = 7.5 Hz, $^4J_{HH}$ = 3.3 Hz, 2 H, C$H_{ar,Fmoc}$), 7.36-7.28 (m, 3 H, C$H_{ar,Fmoc}$, H-5$_{coum}$),), 7.24 (dt, $^3J_{HH}$ = 7.5 Hz, $^4J_{HH}$ = 1.2 Hz, 2 H, C$H_{ar,Fmoc}$), 6.64 (dd, $^3J_{HH}$ = 9.0 Hz, $^4J_{HH}$ = 2.5 Hz, 1 H, H-6$_{coum}$), 6.49 (d, $^4J_{HH}$ = 2.2 Hz, 1 H, H-8$_{coum}$), 6.04 (s, 1 H, H-3$_{coum}$), 5.17 (d, $^4J_{HH}$ = 1.0 Hz, 2 H, CH_2O$_{coum}$), 4.32-4.20 (m, 3 H, α-CH, C$H_{2,Fmoc}$), 4.09 (t, $^3J_{HH}$ = 7.1 Hz, C$H_{sp3,Fmoc}$) 3.35 (q, $^3J_{HH}$ = 7.1 Hz, 4 H, CH_2CH$_3$), 2.53 (m, 2 H, γ-CH_2), 2.29-2.18 (m, 2 H, β-CH_2) 1.09 (t, $^3J_{HH}$ = 7.1 Hz, 6 H, CH$_2$CH_3).

^{13}C-NMR (75 MHz, CDCl$_3$): δ (ppm) = 175.1 (CO$_2$H), 173.5 (γ-CO), 164.0 (C-2$_{coum}$), 158.6 (CO$_{Fmoc\text{-}carbamate}$), 157.3 (C-8a$_{coum}$), 154.3 (C-7$_{coum}$), 145.2 (C-4$_{coum}$), 145.1 ($C_{q,Fmoc}$), 142.5 ($C_{q,Fmoc}$), 128.8 (CH$_{ar,Fmoc}$), 126.3 (CH$_{ar,Fmoc}$), 126.2 (C-5$_{coum}$), 120.9 (CH$_{ar,Fmoc}$), 111.0 (C-3$_{coum}$), 108.1 (C-4a$_{coum}$), 107.0 (C-6$_{coum}$), 99.1 (C-8$_{coum}$), 68.1 (CH$_{2,Fmoc}$), 62.6 (CH$_2$O$_{coum}$), 54.4 (α-CH), 48.4 (CH$_{sp3,Fmoc}$), 46.3 (CH$_2$CH$_3$), 31.2 (γ-CH$_2$), 27.9 (β-CH$_2$), 12.6 (CH$_2$$CH_3$).

ESI-MS m/z: 599.3 [M+H]$^+$, 621.2 [M+Na]$^+$.

HR-MS (ESI): calcd. for C$_{34}$H$_{35}$N$_2$O$_8$, ([M+H]$^+$): 599.2388, found: 599.2384; calcd. for C$_{34}$H$_{34}$N$_2$NaO$_8$, ([M+Na]$^+$): 621.2207, found: 621.2191.

7-(Diethylamino)-4-formylcoumarin

C$_{14}$H$_{17}$NO$_2$
231.3 g/mol
15

C$_{14}$H$_{15}$NO$_3$
245.3 g/mol
34

7-Diethylamino-4-methylcoumarin (4.21 g, 18.2 mmol, 1.0 eq) and selenium dioxide (2.10 g, 18.9 mmol, 1.04 eq) were dissolved in xylene (350 mL) and heated to reflux for 48 h. The completely cooled mixture was filtered and the filtrate was reduced *in vacuo*. The residue was dissolved in DCM (100 mL), adsorbed to silica (10 g) and purified by column chromatography (DCM). The product (2.24 g, 9.1 mmol, 51%) was obtained as a red oily solid.

TLC: R_f = 0.28 (DCM)

^{1}H-NMR (400 MHz, CDCl$_3$): δ (ppm) = 10.0 (s, 1 H, CHO), 8.28 (d, $^3J_{HH}$ = 9.2 Hz, 1 H, H-5$_{coum}$), 6.61 (dd, $^3J_{HH}$ = 9.2 Hz, $^4J_{HH}$ = 2.6 Hz, 1 H, H-6$_{coum}$), 6.50 (d, $^4J_{HH}$ = 2.6 Hz, 1 H, H-8$_{coum}$), 6.43 (s, 1 H, H-3$_{coum}$), 3.41 (q, $^3J_{HH}$ = 7.2 Hz, 4 H, CH_2CH$_3$), 1.21 (t, $^3J_{HH}$ = 7.1 Hz, 6 H, CH$_2$CH_3).

^{13}C-NMR (75 MHz, CDCl$_3$): δ (ppm) = 192.4 (CHO), 161.8 (C-4$_{coum}$), 157.3 (C-2$_{coum}$), 151.0 (C-8a$_{coum}$), 143.8 (C-7$_{coum}$), 128.8 (C-5$_{coum}$), 117.2 (C-3$_{coum}$), 109.5 (C-4a$_{coum}$), 103.6 (C-6$_{coum}$), 97.5 (C-8$_{coum}$), 44.7 (CH$_2$CH$_3$), 12.40 (CH$_2$CH$_3$).

ESI-MS *m/z* : 246.1 [M+H]$^+$, 268.1 [M+Na]$^+$.

HR-MS (ESI): calcd. for C$_{14}$H$_{16}$NO$_3$, ([M+H]$^+$): 246.1127, found: 246.1126; calcd. for C$_{14}$H$_{15}$NNaO$_3$, ([M+Na]$^+$): 268.0946, found: 246.0944.

7-(Diethylamino)-4-(1-hydroxybut-3-en-1-yl)-coumarin

C$_{14}$H$_{15}$NO$_3$
245.3 g/mol
34

C$_{17}$H$_{21}$NO$_3$
287.2 g/mol
33

To a solution of **34** (1.14 g, 4.64 mmol, 1.00 eq) in MeCN/H$_2$O (4:1, 20 mL) ZnCl$_2$ (1.01 g, 7.40 mmol, 1.60 eq) and allyltributylstannane (2.30 mL, 2.46 g, 7.42 mmol, 1.60 eq) were added and the reaction mixture was stirred for 12 h. MeCN was removed under reduced pressure and H$_2$O (20 mL) was added. The aqueous suspension was extracted with DCM (3x20 mL). The combined organic phases were washed with brine (10 mL), dried over MgSO$_4$ and adsorbed onto silica (2 g). Purification by column chromatography (hexane/EtOAc 5:1 to hexane/EtOAc 3:1) gave the product (1.23 g, 4.30 mmol, 93%) as a brown viscous oil.

TLC: R_f = 0.43 (hexane/EtOAc 1:1).

^{1}H-NMR (300 MHz, CDCl$_3$): δ (ppm) = 7.37 (d, $^3J_{HH}$ = 9.1 Hz, 1 H, H-5$_{coum}$), 6.57 (dd, $^3J_{HH}$ = 9.1 Hz, $^4J_{HH}$ = 2.6 Hz, 1 H, H-6$_{coum}$), 6.48 (d, $^4J_{HH}$ = 2.6 Hz, 1 H, H-8$_{coum}$), 6.24 (t, $^4J_{HH}$ = 1.3 Hz, 1 H, H-3$_{coum}$), 5.95-5.80 (m, 1 H, CH_{allyl}), 5.24-5.15 (m, 2 H, $CH_{2,sp2,allyl}$), 5.01 (dd, $^3J_{HH}$ = 7.89 Hz, $^3J_{HH}$ = 3.7 Hz, $CHOH_{coum}$), 3.39 (q, J = 7.1 Hz, 4 H, CH_2CH_3), 2.71-2.38 (m, 2 H, $CH_{2,sp3,allyl}$), 1.19 (t, J = 7.1 Hz, 6 H, CH_2CH_3).

^{13}C-NMR (75 MHz, CDCl$_3$): δ (ppm) = 162.4 (C-2$_{coum}$), 157.3 (C-8a$_{coum}$), 156.3 (C-4$_{coum}$), 150.2 (C-7$_{coum}$), 133.2 (CH_{allyl}), 124.8 (C-5$_{coum}$), 119.1, $CH_{2,sp2,allyl}$), 108.5 (C-6$_{coum}$), 106.1 (C-4a$_{coum}$), 105.4 (C-3$_{coum}$), 98.0 (C-8$_{coum}$), 68.8 ($CHOH_{coum}$), 44.7 (CH_2CH_3), 12.5 (CH_2CH_3).

ESI-MS m/z: 288.2 [M+H]$^+$, 310.2 [M+Na]$^+$, 575.3 [2M+H]$^+$.

HR-MS (ESI): calc. for C$_{17}$H$_{22}$NO$_3$, ([M+H]$^+$): 288.1594, found: 288.1594; calc. for C$_{17}$H$_{21}$NNaO$_3$, ([M+Na]$^+$): 310.1414, found: 310.1420.

Fmoc-L-Glu(DEACM^{allyl})-O^{t}Bu

C$_{17}$H$_{21}$NO$_3$
287.2 g/mol
33

C$_{41}$H$_{46}$N$_2$O$_8$
694.8 g/mol
91

To a solution of **33** (0.60 g, 2.09 mmol, 1.00 eq) and Fmoc-L-Glu-O^{t}Bu (0.98 g, 2.30 mmol, 1.10 eq) in anhydrous DCM (50 mL) DCC (0.47 g, 2.30 mmol, 1.10 eq) and

DMAP (26 mg, 0.21 mmol, 0.10 eq) were added. The solution was stirred for 12 h at rt. Then, the reaction mixture was washed with saturated NaHCO$_3$ solution (3 x 50 mL), dried over MgSO4, and adsorbed onto silica (3 g). Purification by column chromatography (DCM/MeOH, 95:5) yielded the product (1.03 g, 1.93 mmol, 96%) as a yellow solid.

TLC: R_f = 0.34 (DCM/MeOH 95:5).

^{1}H-NMR (400 MHz, CDCl$_3$): δ (ppm) = 7.76 (d, $^3J_{HH}$ = 7.5 Hz, 2 H, C$H_{ar,Fmoc}$), 7.60 (d, $^3J_{HH}$ = 7.6 Hz, 2 H, C$H_{ar,Fmoc}$), 7.43-7.37 (m, 3 H, H-5$_{coum}$, C$H_{ar,Fmoc}$), 7.34-7.28 (m, 2 H, C$H_{ar,Fmoc}$), 6.60-6.55 (m, 1 H, H-6$_{coum}$), 6.51-6.49 (m, 1 H, H-8$_{coum}$), 6.09-6-01 (m, 2 H, CHO$_{coum}$, H-3$_{coum}$), 5.81-5.69 (m, 1 H, C$H_{sp2,allyl}$), 5.14-5.07 (m, 2 H, C$H_{2,sp2,allyl}$), 4.43-4.27 (m, 3 H, α-CH, C$H_{2,Fmoc}$), 4.25-4.18 (m, 1 H, C$H_{sp3,Fmoc}$), 3.39 (q, $^3J_{HH}$ = 7.0 Hz, 4 H, CH_2CH$_3$), 2.72-2.38 (m, 4 H, C$H_{2,sp3,allyl}$, γ-CH_2), 1.98-1.64 (m, 2 H, β-CH_2), 1.47 (s, 3 H, C$H_{3,tBu}$) 1.19 (t, $^3J_{HH}$ = 7.2 Hz, 6 H, CH$_2$CH_3).

^{13}C-NMR (100 MHz, CDCl$_3$): δ (ppm) = 171.8 (CO$_{2,tBu}$), 171.0 (γ-CO), 162.1 (C-2$_{coum}$), 156.7 (CO$_{Fmoc-carbamate}$), 157.8 (C-8a$_{coum}$), 153.7 (C-4$_{coum}$), 150.7 (C-7$_{coum}$), 143.9 ($C_{q,Fmoc}$), 141.4 ($C_{q,Fmoc}$), 132.3 (CH$_{sp2,allyl}$), 127.8 (CH$_{ar,Fmoc}$), 127.2 (CH$_{ar,Fmoc}$), 125.3 (C-5$_{coum}$), 124.9 (CH$_{ar,Fmoc}$), 120.1 (CH$_{ar,Fmoc}$), 119.1 (CH$_{2,sp2,allyl}$), 108.9 (C-6$_{coum}$), 106.0 (C-4a$_{coum}$), 105.8 (C-3$_{coum}$), 98.2 (C-8$_{coum}$), 82.1 ($C_{q,tBu}$) 70.6 (CHO$_{coum}$), 67.2 (CH$_{2,Fmoc}$), 53.8 (α-CH), 47.3 (CH$_{sp3,Fmoc}$), 44.9 (CH$_2$CH$_3$), 39.2 (CH$_{2,sp3,allyl}$), 32.1 (γ-CH$_2$), 28.2 (CH$_{3,tBu}$), 28.1 (β-CH$_2$), 12.6 (CH$_2$CH$_3$).

ESI-MS m/z : 695.3 [M+H]$^+$, 717.3 [M+Na]$^+$.

HR-MS (ESI): calcd. for C$_{41}$H$_{47}$N$_2$O$_8$, ([M+H]$^+$): 695.3327, found: 639.3325; calcd. for C$_{41}$H$_{46}$N$_2$NaO$_8$, ([M+Na]$^+$): 717.3136, found: 717.3136.

Fmoc-L-Glu(DEACM^{allyl})-OH

C$_{41}$H$_{46}$N$_2$O$_8$
694.8 g/mol
91

C$_{37}$H$_{38}$N$_2$O$_8$
638.7 g/mol
38

91 (1.37 g, 1.97 mmol, 1.0 eq) was dissolved in DCM/TFA (1:1, 60 mL) and stirred at rt for 1 h. The solvents were removed in a N$_2$ stream, residual TFA was coevaporated under reduced pressure with DCM (3 x 50 mL). The product (1.26 g, 1.97 mmol, quant.) was obtained as a yellow solid.

TLC: R_f = 0.34 (DCM/MeOH/AcOH 95:5:0.1).

^{1}H-NMR (600 MHz, CD$_3$OD): δ (ppm) = 7.76 (d, $^3J_{HH}$ = 7.6 Hz, 2 H, CH_{Fmoc}), 7.62 (d, $^3J_{HH}$ = 7.6 Hz, 2 H, CH_{Fmoc}), 7.52 (d, $^3J_{HH}$ = 9.2 Hz, 1 H, H-5$_{coum}$), 7.38-7.33 (m, 2 H, CH_{Fmoc}), 7.31-7.25 (m, 2 H, CH_{Fmoc}), 6.68 (dd, $^3J_{HH}$ = 9.1 Hz, $^4J_{HH}$ = 2.6 Hz, 1 H, H-6$_{coum}$), 6.47 (d, $^4J_{HH}$ = 2.4 Hz, 1 H, H-8$_{coum}$), 6.11-6.07 (m, 1 H, CHO$_{coum}$), 6.02 (s, 1 H, H-3$_{coum}$), 5.86-5.76 (m, 1 H, C$H_{sp2,allyl}$), 5.12-5.04 (m, 2 H, C$H_{2,sp2,allyl}$), 4.37-4.22 (m, 3 H, α-CH, C$H_{2,Fmoc}$), 4.19-4.13 (m, 1 H, CH$_{sp3,Fmoc}$), 3.41 (q, $^3J_{HH}$ = 7.2 Hz, 4 H, CH_2CH$_3$), 2.71-2.51 (m, 4 H, C$H_{2,sp3,allyl}$, γ-CH_2), 2.28-1.92 (m, 2 H, β-CH_2), 1.16 (t, $^3J_{HH}$ = 7.2 Hz, 6 H, CH$_2$CH_3).

^{13}C-NMR (75 MHz, CD$_3$OD): δ (ppm) = 174.4 (CO_2H), 173.0 (γ-CO), 164.2 (C-2$_{coum}$), 158.4 ($CO_{Fmoc\text{-}carbamate}$), 157.6 ($C$-8a$_{coum}$), 156.5 ($C-4_{coum}$), 152.2 ($C-7_{coum}$), 145.1 ($C_{q,Fmoc}$), 142.4 ($C_{q,Fmoc}$), 133.5 ($CH_{sp2,allyl}$), 128.7 ($CH_{ar,Fmoc}$), 128.0 ($CH_{ar,Fmoc}$), 126.4 ($C-5_{coum}$), 126.3 ($CH_{ar,Fmoc}$), 120.8 ($CH_{ar,Fmoc}$), 119.2 ($CH_{2,sp2,allyl}$), 110.5 ($C-6_{coum}$), 106.9 ($C$-4a$_{coum}$), 105.5 ($C-3_{coum}$), 98.5 ($C-8_{coum}$), 71.9 ($CHO_{coum}$), 67.9 ($CH_{2,Fmoc}$), 54.4 ($\alpha$-$CH$), 48.4 ($CH_{sp3,Fmoc}$), 45.6 ($CH_2CH_3$), 40.1 ($CH_{2,sp3,allyl}$), 31.4 ($\gamma$-$CH_2$), 27.8 ($\beta$-$CH_2$), 12.8 ($CH_2$$CH_3$).

ESI-MS m/z: 639.4 [M+H]$^+$, 661.3 [M+Na]$^+$.

HR-MS (ESI): calcd. for C$_{37}$H$_{39}$N$_2$O$_8$, ([M+H]$^+$): 639.2701, found: 639.2694; calcd. for C$_{37}$H$_{38}$N$_2$NaO$_8$, ([M+Na]$^+$): 661.2520, found: 661.2506.

1-(7-Diethylaminocoumarin-4-yl)but-3-en-1-yl (4-nitrophenyl) carbonate

HO

O$_2$N

C$_{17}$H$_{21}$NO$_3$
287.2 g/mol
33

C$_{24}$H$_{24}$N$_2$O$_7$
452.5 g/mol
36

33 (1.08 g, 3.76 mmol, 1.00 eq) and 4-nitrophenyl chloroformate (1.14 g, 5.64 mmol, 1.50 eq) were suspended in DCM (25 mL), cooled with an ice bath, and DIPEA (0.65 mL, 3.76 mmol, 1.00 eq) was added yielding a clear solution. The solution was stirred for 12 h, allowing it to come to room temperature gradually. The newly formed precipitate was dissolved by addition of DIPEA (0.65 mL, 4.0 mmol, 1.0 eq) and the solution was stirred for 1 h. The reaction mixture was directly adsorbed on silica and the crude product was purified by column chromatography (gradient DCM to DCM/acetone 10:1). The product (0.544 g, 1.20 mmol, 32%) was obtained as a yellow solid.

TCL: $R_f = 0.16$ (DCM)

^{1}H-NMR (500 MHz, CDCl$_3$): δ (ppm) = 8.30 (dt, $^3J_{HH}$ = 9.2 Hz, $^4J_{HH}$ = 2.2 Hz, 2 H, H-3$_{pNP}$, H-5$_{pNP}$), 7.42 (d, $^3J_{HH}$ = 9.1 Hz, 1 H, H-5$_{coum}$), 7.38 (dt, $^3J_{HH}$ = 9.2 Hz, $^5J_{HH}$ = 2.2 Hz, 2 H, H-5$_{pNP}$, H-6$_{pNP}$), 6.60 (dd, $^3J_{HH}$ = 9.1 Hz, $^4J_{HH}$ = 2.6 Hz, 1 H, H-6$_{coum}$), 6.51 (d, $^4J_{HH}$ = 2.6 Hz, 1 H, H-8$_{coum}$), 6.17 (s, 1 H, H-3$_{coum}$), 5.95 (dd, $^3J_{HH}$ = 7.6 Hz, $^3J_{HH}$ = 4.8 Hz, 1 H, CHO$_{coum}$), 5.88-5.78 (m, 1 H, C$H$$_{sp2,allyl}$), 5.24-5.17 (m, 2 H, C$H_2$$_{sp2,allyl}$), 3.40 (q, $^3J_{HH}$ = 7.1 Hz, 4 H, CH_2CH$_3$), 2.80-2.67 (m, 2 H, C$H_2$$_{sp3,allyl}$), 1.19 (t, $^3J_{HH}$ = 7.1 Hz, 6 H, CH$_2$CH_3).

^{13}C-NMR (125 MHz, CDCl$_3$): δ (ppm) = 161.8 (C-2$_{coum}$), 156.6 (C-8a$_{coum}$), 155.2 (C-1$_{pNP}$), 152.4 (C-4$_{coum}$), 151.7 (CO$_{carbonate}$), 150.7 (C-7$_{coum}$), 145.5 (C-4$_{pNP}$), 131.6 (allyl$_{sp2}$-CH), 125.3 (C-3$_{pNP}$, C-5$_{pNP}$), 124.6 (C-5$_{coum}$), 121.7 (C-2$_{pNP}$, C-6$_{pNP}$), 119.8 (CH$_2$$_{sp2,allyl}$), 108.9 ($C-6_{coum}$), 105.6 ($C-3_{coum}$), 105.4 ($C$-4a$_{coum}$), 98.1 ($C-8_{coum}$), 75.6 (CHO$_{coum}$), 44.9 (CH$_2CH_3$), 39.0 (CH$_2$$_{sp3,allyl}$), 12.4 (CH$_2CH_3$).

ESI-MS m/z: 453.2 [M+H]$^+$, 475.2 [M+Na]$^+$.

HR-MS (ESI): calcd. for C$_{24}$H$_{25}$N$_2$O$_7$, ([M+H]$^+$): 453.1656, found: 453.1655; calcd. for C$_{24}$H$_{24}$N$_2$NaO$_7$, ([M+Na]$^+$): 475.1476, found: 475.1477.

Fmoc-L-Lys(DEACM^{allyl})-OH

C$_{24}$H$_{24}$N$_2$O$_7$
452.5 g/mol
36

C$_{39}$H$_{43}$N$_3$O$_8$
681.8 g/mol
37

36 (0.54 g, 1.24 mmol, 1.00 eq) suspended in DMF/DCM 1:1 (20 mL) was combined with Fmoc-L-Lys-OH (0.46 g, 1.24 mmol, 1.00 eq) suspended in toluene/DCM 3:2 (30 mL) and cooled in an ice bath. DIPEA (216 µL, 1.24 mmol, 1.00 eq) was added and the reaction mixture was stirred for 12 h allowing it to slowly come to room temperature. Further DIPEA (50 µL, 0.29 mmol, 0.23 eq) was added and the reaction mixture was stirred at rt for 1 h. The solvents were removed under reduced pressure. The residue was dissolved in DCM (50 mL), adsorbed to silica (3 g), and purified by column chromatography (DCM/MeOH/AcOH, 95:5:0.1). The product (178 mg, 1.70 mmol, 22%) was obtained as a yellow powder.

TLC: R_f = 0.29 (DCM/MeOH/AcOH 95:5:0.1).

^{1}H-NMR (400 MHz, CDCl$_3$): δ (ppm) = 7.72 (d, $^3J_{HH}$ = 7.5 Hz, 2 H, C$H_{ar,Fmoc}$), 7.62-7.52 (m, 2 H, C$H_{ar,Fmoc}$), 7.41-7.32 (m, 3 H, H-5$_{coum}$, C$H_{ar,Fmoc}$), 7.31-7.24 (t, $^3J_{HH}$ = 7.5 Hz, 2 H, C$H_{ar,Fmoc}$), 6.64-6.52 (m, 1 H, H-6$_{coum}$), 6.52-6.47 (m, 1 H, H-8$_{coum}$), 6.10 (s, 1 H, H-3$_{coum}$), 5.95-5.87 (m, 1 H, CHO$_{coum}$), 5.79-5.72 (m, 1 H, C$H_{sp2,allyl}$), 5.14-5.03 (m, 2 H, C$H_{2,sp2,allyl}$), 4.44-4.29 (m, 3 H, α-CH, C$H_{2,Fmoc}$), 4.21-4.14 (m, 1 H, C$H_{sp3,Fmoc}$), 3.38 (q, $^3J_{HH}$ = 7.0 Hz, 4 H, CH_2CH$_3$), 3.30-2.96 (m, 2 H, δ-CH_2), 2.70-2.40 (m, 2 H, C$H_{2,sp3,allyl}$), 1.96-1.71 (m, 2 H, β-CH_2), 1.64-1.27 (m, 4 H, δ-CH_2, δ-CH_2), 1.16 (t, $^3J_{HH}$ = 7.2 Hz, 6 H, CH$_2$CH_3).

^{13}C-NMR (100 MHz, CDCl$_3$): δ (ppm) = 173.7 (CO_2H), 163.9 (C-2$_{coum}$), 156.4 (CO$_{coum-carbamate}$), 156.2 (C-8a$_{coum}$), 155.9 (CO$_{Fmoc-carbamate}$), 155.3 (C-7$_{coum}$), 150.6 (C-4$_{coum}$), 143.8 ($C_{q,Fmoc}$), 141.3 ($C_{q,Fmoc}$), 132.4 (C$H_{sp2,allyl}$), 127.7 (C$H_{ar,Fmoc}$), 127.1 (C$H_{ar,Fmoc}$), 125.2 (C-5$_{coum}$), 124.8 (C$H_{ar,Fmoc}$), 119.9 (C$H_{ar,Fmoc}$), 118.5 (C$H_{2,sp2,allyl}$), 109.2 (C-3$_{coum}$), 106.1 (C-6$_{coum}$), 104.0 (C-4a$_{coum}$), 98.2 (C-8$_{coum}$), 70.4 (CHO$_{coum}$), 67.0 (C$H_{2,Fmoc}$), 53.6 (α-CH), 47.1 (Fmoc$_{sp3}$-CH), 44.9 (CH_2CH$_3$), 40.4 (allyl$_{sp3}$-CH_2), 39.0 (ε-CH_2), 32.0 (β-CH_2), 28.9 (β-CH_2), 21.6 (γ-CH_2), 12.6 (CH$_2$CH_3).

ESI-MS m/z: 639.4 [M+H]$^+$, 661.3 [M+Na]$^+$.

HR-MS (ESI): calcd. for C$_{37}$H$_{39}$N$_2$O$_8$, ([M+H]$^+$): 639.2701, found: 639.2694; calcd. for C$_{37}$H$_{38}$N$_2$NaO$_8$, ([M+Na]$^+$): 661.2520, found: 661.2506.

H-L-Lys(DEACM^{allyl})-OH

C$_{39}$H$_{43}$N$_3$O$_8$
681.8 g/mol
37

C$_{24}$H$_{33}$N$_3$O$_6$
459.5 g/mol
39

37 (100 mg, 0.14 mmol, 1.00 eq) was dissolved in piperidin/DMF (1:4, 10 mL) and stirred at rt for 20 min. The solvents were removed under reduced pressure. The residue was suspended in H$_2$O (30 mL), centrifuged (5 min, 9000 rt/min, 20 °C) and the supernatant was lyophilized. After purification by HPLC and lyophilization the product (46 mg, 0.10 mmol, 71%) was obtained as a fine yellow powder.

HPLC: R_t = 9.15 min.

100

¹H-NMR (400 MHz, CD₃OD): δ (ppm) = 7.59 (d, $^3J_{HH}$ = 9.2 Hz, 2 H, *H*-5$_{coum}$), 6.78 (dd, $^3J_{HH}$ = 9.2 Hz, $^4J_{HH}$ = 2.6 Hz, 1 H, *H*-6$_{coum}$), 6.55 (d, $^3J_{HH}$ = 2.6 Hz, 1 H, *H*-8$_{coum}$), 6.03 (s, 1 H, *H*-3$_{coum}$), 5.98-5.92 (m, 1 H, C*H*O$_{coum}$), 5.91-5.80 (m, 1 H, C*H*$_{sp2,allyl}$), 5.16-5.08 (m, 2 H, C*H*₂$_{,sp2,allyl}$), 3.84-3.77 (m, 1 H, α-C*H*), 3.49 (q, $^3J_{HH}$ = 7.1 Hz, 4 H, C*H*₂CH₃), 3.21-3.07 (m, 2 H, ε-C*H*₂), 2.74-2.54 (m, 2 H, C*H*₂$_{,sp3,allyl}$), 2.02-1.77 (m, 2 H, β-C*H*₂), 1.64-1.27 (m, 4 H, γ-C*H*₂, δ-C*H*₂), 1.16 (t, $^3J_{HH}$ = 7.2 Hz, 6 H, CH₂C*H*₃).

¹³C-NMR (100 MHz, CD₃OD): δ (ppm) = 173.7 (*C*O₂H), 164.8 (*C*-2$_{coum}$), 158.1 (*C*-8a$_{coum}$) 157.8 (*C*O$_{carbamat}$), 157.6 (*C*-4$_{coum}$), 152.5 (*C*-7$_{coum}$), 133.9 (*C*H$_{sp2,allyl}$), 126.5 (*C*-5$_{coum}$), 118.9 (*C*H₂$_{,sp2,allyl}$), 110.6 (*C*-6$_{coum}$), 106.9 (*C*-4a$_{coum}$), 104.7 (*C*-3$_{coum}$), 98.4 (*C*-8$_{coum}$), 72.2 (*C*HO$_{coum}$), 54.6 (α-*C*H), 45.6 (*C*H₂CH₃), 41.3 (ε-*C*H₂), 40.3 (allyl$_{sp3}$-*C*H₂), 31.4 (β-*C*H₂), 30.4 (δ-*C*H₂), 23.4 (γ-*C*H₂), 17.7 (*C*H₂*C*H₃).

ESI-MS *m/z*: 460.2 [M+H]⁺, 482.2 [M+Na]⁺, 919.5 [2M+H]⁺, 941.5 [2M+Na]⁺.

HR-MS (ESI): calcd. for C₂₄H₃₄N₃O₆, ([M+H]⁺): 460.2441, found: 460.2442; calcd. for C₂₄H₃₃N₃NaO₆, ([M+Na]⁺): 482.2267, found: 482.2262.

1,4-Di(H-L-Lys(DEACM)-OH)but-2-en

C₃₉H₄₃N₃O₈
681.8 g/mol
37

C₄₆H₆₂N₆O₁₂
891.0 g/mol
40

37 (150 mg, 0.22 mmol, 1.00 eq) and Grubbs I catalyst (10 mg) dissolved in anhydrous DCM (15 mL) were heated to reflux for 8 h. The solvent was removed under reduced pressure. The residue was dissolved in piperidin/DMF (1:4) and stirred at rt for 20 min. The solvents were removed under reduced pressure. The residue was suspended in H₂O (30 mL), centrifuged (5 min, 9000 rt/min, 20 °C) and the supernatant was lyophilized. After purification by HPLC and lyophilization the product (53 mg, 59 μmol, 54%) was obtained as a fine yellow powder. For NMR analysis only the *cis*-isomers were isolated.

HPLC: R_t = 9.87 min (cis), 10.06 min (trans), 10.21 min (trans).

^{1}H-NMR (400 MHz, CD$_3$OD): δ (ppm) = 7.49 (d, ^3J = 9.2 Hz, 2 H, *H*-5$_{coum}$), 6.69 (dd, ^3J = 9.2 Hz, ^4J = 2.6 Hz, 1 H, *H*-6$_{coum}$), 6.52 (d, ^3J = 2.6 Hz, 1 H, *H*-8$_{coum}$), 5.93 (d, ^4J =1.44 Hz, 1 H, *H*-3$_{coum}$), 5.89-5.83 (m, 1 H, C*H*O$_{coum}$), 5.60 (t, $^3J_{HH}$ = 4.8 Hz, 1 H, C*H*$_{butenyl}$), 3.91 (q, ^3J = 5.5 Hz, 1 H, α-C*H*), 3.45 (q, ^3J = 7.4 Hz, 4 H, C*H*$_2$CH$_3$), 3.20-3.04 (m, 2 H, ε-C*H*$_2$), 2.62-2.37 (m, 2 H, C*H*$_{2,butenyl}$), 2.00-1.79 (m, 2 H, β-C*H*$_2$), 1.63-1.37 (m, 4 H, γ-C*H*$_2$, δ-C*H*$_2$), 1.13 (t, ^3J = 7.1 Hz, 6 H, CH$_2$C*H*$_3$).

^{13}C-NMR (100 MHz, CD$_3$OD): δ (ppm) = 171.9 (*C*O$_2$H), 164.8 (*C*-2$_{coum}$), 157.8 (*C*-8a$_{coum}$) 157.7 (*C*O$_{carbamat}$), 157.6 (*C*-4$_{coum}$), 152.4 (*C*-7$_{coum}$), 128.3 (*C*H$_{butenyl}$), 126.5 (*C*-5$_{coum}$), 110.6 (*C*-6$_{coum}$), 106.8 (*C*-4a$_{coum}$), 104.7 (*C*-3$_{coum}$), 98.5 (*C*-8$_{coum}$), 72.3 (*C*HO$_{coum}$), 53.9 (α-*C*H), 45.6 (*C*H$_2$CH$_3$), 41.3 (ε-*C*H$_2$), 33.7 (*C*H$_{2,butenyl}$), 31.2 (β-*C*H$_2$), 30.3 (δ-*C*H$_2$), 23.3 (γ-*C*H$_2$), 12.7 (CH$_2$*C*H$_3$).

ESI-MS *m/z*: 891.5 [M+H]$^+$, 446.2 [M+2H]$^{2+}$.

HR-MS (ESI): calcd. for C$_{46}$H$_{63}$N$_6$O$_{12}$, ([M+H]$^+$): 891.4499, found: 891.4498; calcd. for C$_{46}$H$_{64}$N$_6$O$_{12,}$ ([M+2H]$^{2+}$): 446.2282, found: 446.2286.

7.6 Peptide synthesis and related reactions

7.6.1 Solid phase peptide synthesis (SPPS)

Peptides were synthesized on a on a pre-loaded *Wang* resin (0.27-0.32 mmol/g, *Nova Biochem* (Darmstadt, Germany)) for carboxy *C*-termini or a Rink amide resin (0.5 mmol/g) for amide *C*-termini. The following commercially available L-amino acid (aa) building blocks were used in automated microwave assisted SPPS (see below):

Fmoc-Ala-OH (A)	Fmoc-Ile-OH (I)	Fmoc-Thr(*t*Bu)-OH (T)
Fmoc-Arg(Pbf)-OH (R)	Fmoc-Leu-OH (L)	Fmoc-Trp(Boc)-OH (W)
Fmoc-Asn(Trt)-OH (N)	Fmoc-Lys(Boc)-OH (K)	Fmoc-Tyr(*t*Bu)-OH (Y)
Fmoc-Cys(Trt)-OH (C)	Fmoc-Met-OH (M)	Fmoc-Val-OH (V)
Fmoc-Gln(Trt)-OH (Q)	Fmoc-Phe-OH (F)	Fmoc-Pra-OH ({pra})
Fmoc-Glu(OtBu)-OH (E)	Fmoc-Pro-OH (P)	Fmoc-AHA-OH ({az})
Fmoc-Gly-OH (G)	Fmoc-Ser(tBu)-OH (S)	

Building blocks synthesized for this work were coupled by manual microwave assisted SPPS (see below) after transfer of the resin from the synthesizer reaction vessel to a *Discardit II* syringe by *Becton Dickinson* (Heidelberg, Germany) equipped with a polyethylene frit, in the following referred to as BD syringe. After coupling, the excess building block was recovered by quenching and precipitation with water. Centrifugation and purification of the pallet by chromatography yielded an average of 2 of 5 utilized equivalents.

Fmoc-L-Lys(NVOC)-OH	Fmoc-L-Glu(DMNPB)-OH
Fmoc-L-Lys(DEACM)-OH	Fmoc-L-Glu(DEACM)-OH
Fmoc-L-Lys(DEACM^{allyl})-OH	Fmoc-L-Glu(DEACM^{allyl})-OH

Furthermore, when opportune for the following of coupling efficiency or side reactions, commercially available amino acids were also coupled by manual SPPS. Commonly repeated processes are successively summarized as standard operating procedures (**SOP**s).

7.6.2 Automated solid phase peptide synthesis

Automated SPPS was performed 0.05 mmol scale on a *Liberty Blue CEM* (Matthews, USA) microwave assisted peptide synthesizer. **SOP1**: For peptides of up to 30 aas length, the synthesis was conducted via a standard Fmoc/tBu-protocol using the recommended (single) coupling (5 eq. aa, 5 eq. DIC, 5 eq. OxymaPure and 0.5 eq. DIPEA in DMF, 1: 75 °C, 170 W, 15 s, 2: 90 °C, 30 W, 105 s) and deprotection methods (piperidine/DMF, 1:4, *v/v*, 1: 75 °C, 155W, 15 s, 2: 90 °C, 30W, 50 s). **SOP2**: Peptides of more than 30 aas length were coupled twice per aa using the modified CarbomaMAX™ coupling (5 eq. aa, 10 eq. DIC, 5 eq. OxymaPure, 0.5 eq. DIPEA in DMF, 1: 75 °C, 170 W, 15 s, 2: 90 °C, 30 W, 105 s) method. Special care was taken for the incorporation of Cys, and Arg residues. For cysteine, the temperature of the microwave assisted coupling was reduced and the reaction time elongated (1: 25 °C, 0 W, 120 s, 2: 50 °C, 30 W, 480 s). Arginine was always introduced by double coupling ((a) 1: 25 °C, 0 W, 1500 s, 2: 75 °C, 30 W, 120 s, (b) 1: 75 °C, 30 W, 300 s).

7.6.3 Manual SPPS

Manual coupling was performed using a *Discover* microwave reaction cavity by *CEM* (Matthews, USA). The resin was placed in a BD syringe. **SOP3**: Double coupling of caged aas was achieved by treatment with the coupling cocktail (5 eq. aa, 5 eq. HATU, 4.5 eq, HOAt, 10 eq. DIPEA in DMF) and supported by microwave irradiation (50 °C, 25 W, 10 min). **SOP4**: Standard aas were introduced at higher temperatures (75 °C, 30 W, 5 min). Fmoc was removed with piperidine (20% in DMF, 1: 50 °C, 35 W, 30 s, 2: 75 °C, 25 W, 180 s). Between all steps, the resin was washed (5 x DMF).

7.6.4 *N*-terminal acylation

SOP5: To obtain an uncharged *N*-terminus the resin bound peptide was treated with the acylation cocktail (10% Ac$_2$O, 5% DIPEA in DMF) for 10 min at room temperature. The process was repeated twice.

7.6.5 On resin metathesis

SOP6: Macrocyclization of the *N*-terminally protected or acylated peptides was performed on resin. After SPPS the resin was washed successively with DMF (5 x), MeOH (5 x), and DCM (5 x) and dried in a desiccator. A two neck round bottom flask equipped with a reflux condenser containing the resin (10-100 mg) and the catalyst (0.25 mg/ mg of resin) were carefully purged with argon using the SCHLENK technique. Degassed DCE (2 mL/mg resin) was added and the mixture was stirred at 50 °C for 2 h. The resin was filtered off and the process was repeated with fresh catalyst and solvent to complete the reaction. After filtration and thorough washing of the resin with DCE (10 x) and DMF (10 x), Fmoc deprotection (20% piperidine in DMF, 2 x 10 min at room temperature) was performed if necessary.

7.6.6 Cleavage

SOP7: After synthesis, the resin was filtered off, washed successively with DMF (5 x), MeOH (5 x) and DCM (10 x) and dried *in vacuo*. Acidic cleavage from the resin was achieved by treatment with a mixture of a) trifluoroacetic acid (TFA)/triisopropylsilane/water (95:2.5:2.5, 5 mL, 3 h, cleavage) for sequences without cysteine or b) TFA/triisopropylsilane/ethanedithiol/water (94:2.5:2.5:1, 5 mL, 3 h) for cysteine containing peptides. The resin was extracted with additional TFA (5 mL) and the combined extracts were concentrated to 2 mL under a flow of nitrogen. The crude peptide was then precipitated in cold diethyl ether (30 mL) and isolated by centrifugation and decantation of the supernatant. The precipitate was washed twice with ice-cold diethyl ether and subsequently lyophilized.

For test cleavage usually a small spatula-tipfull of resin was placed in a 2 mL BD syringe and treated as described above only with adjusted volumes (300 µL cleavage cocktail, 1 mL diethyl ether). After lyophilization the peptide was analyzed by UPLC and mass spectrometry.

7.6.7 KAISER test

For qualitative verification of coupling completion, the KAISER test was carried out. The test detects primary amino groups. A small amount of resin was placed in a screw top glass and 50 µL of each of the following solutions was added:

A: ninhydrin (5 g) in ethanol (100 mL)

B: phenol (80 g) in ethanol (20 mL)

C: KCN (1 mM in water, 2 mL) in pyridine (96 mL)

The sample was incubated at 100 °C for 5 min. The presence of primary amino groups and thus, incomplete coupling was verified by a deep blue color of the solution and the resin beads.

DEACM-test-peptide a

$C_{51}H_{80}N_{12}O_{17}$
1133.3 g/mol
18

The peptide was synthesized at a 10 µmol scale by manual SPPS (**SOP3** and **SOP4**). After cleavage (**SOP7**) and lyophilization the crude product was purified by semi-preparative HPLC utilizing column 3 with a linear gradient of eluent system V (10-80% B in 30 min).

UPLC (column 1, Eluent system II, 10-95% in 20 min) : R_t = 13.0 min.

ESI-MS *m/z*: 1133.6 [M+H]$^+$, 578.3 [M+H+Na]$^{2+}$.

HR-MS (ESI): calcd. for $C_{51}H_{81}N_{12}O_{17}$, ([M+H]$^+$): 1133.5837, found: 1133.5822; calcd. for $C_{51}H_{81}N_{12}NaO_{17}$, ([M+H+Na]$^{2+}$): 578.2865, found: 578.2861.

DEACM test-peptide b

H-W W G K I A A L K E K I A A L K E K I A A L K E Q S K A R R K K G-OH

$C_{187}H_{310}N_{52}O_{46}$
4022.9
19

The peptide was synthesized at a 50 µmolar scale by automated SPPS (**SOP2**) following the protocol for peptides longer than 30 aas. Fmoc-Lys(DEACM)-OH was coupled by manual SPPS (**SOP3**). After cleavage (**SOP7**) and lyophilization the crude product was purified by preparative HPLC utilizing column 5 with a linear gradient of eluent system V (60-80% B in 30 min).

UPLC (column 1, Eluent system II, 20-95% in 20 min) : R_t = 15.45 min.

ESI-MS m/z: 805.5 $[M+5H]^{5+}$, 671.4 $[M+6H+]^{6+}$.

HR-MS (ESI): calcd. for $C_{187}H_{315}N_{52}O_{46}$, ($[M+5H]^+$): 805.0776, found: 805.0784; calcd. for $C_{187}H_{316}N_{52}O_{46}$, ($[M+7H]^{7+}$): 671.0659, found: 671.0661.

Test-peptide c

H-W W G K I A A L K E K I A A L K E K I A A L K E Q S K A R R K K G-OH

$C_{172}H_{295}N_{51}O_{42}$
3749.6 g/mol
20

The peptide was synthesized at a 50 µmolar scale by automated SPPS (**SOP2**) following the protocol for peptides longer than 30 aas. Fmoc-Lys(DEACM)-OH was coupled by manual SPPS (**SOP3**). After cleavage (**SOP7**) and lyophilization the crude product was purified by preparative HPLC utilizing column 5 with a linear gradient of eluent system V (60-80% B in 30 min).

UPLC (column 1, Eluent system II, 20-95% in 20 min) : R_t = 13.3 min.

ESI-MS m/z: 750.9 $[M+5H]^{5+}$, 625.9 $[M+6H+]^{6+}$.

HR-MS (ESI): calcd. for $C_{172}H_{300}N_{51}O_{42}$, ($[M+5H]^{5+}$): 750.4576, found: 750.4569; calcd. for $C_{172}H_{301}N_{51}O_{42}$, ($[M+6H]^{6+}$): 625.5492, found: 625.5483.

Test-peptide d

H-G E I A A L E K E I A A L E K E I A A L E K Y W W K N L K G-OH

$C_{171}H_{275}N_{43}O_{48}$
3701.3 g/mol
21

The peptide was synthesized at a 50 µmolar scale by automated SPPS (**SOP2**) following the protocol for peptides longer than 30 aas. After cleavage (**SOP7**) and lyophilization the crude product was purified by preparative HPLC utilizing column 5 with a linear gradient of eluent system V (50-90% B in 30 min).

UPLC (column 1, Eluent system II, 20-95% in 20 min) : R_t = 16.20 min.

ESI-MS m/z: 1234.7 $[M+3H]^{3+}$, 926.3 $[M+3H]^{4+}$.

HR-MS (ESI): calcd. for $C_{171}H_{278}N_{43}O_{48}$, ($[M+3H]^{3+}$): 1234.0206, found: 1234.0200; calcd. for $C_{171}H_{279}N_{43}O_{48}$, ($[M+6H]^{6+}$): 925.7673, found: 925.7675.

E3((DEACM)$_2$^{butenyl}) a

$C_{140}H_{219}N_{29}O_{40}$
2948.5 g/mol
31

The peptide was synthesized at a 25 µmolar scale on a Rink amid resin. The first 16 aas were coupled by automated SPPS (**SOP1**). All following aas were added by manual SPPS (**SOP3** and **SOP4**) and the reactions around the caged aa were followed by KAISER test and test cleavages. The final Fmoc deprotection was carried out and the N-terminus was acylated (**SOP5**). After thorough washing of the resin Grubbs-catalyzed macrocyclization was performed (**SOP6**). The peptide was cleaved (**SOP7**) and after lyophilization the crude product was purified by a two-step semi-preparative HPLC utilizing column 3. First, truncated peptides from pyroglutamate formation were separated with a linear gradient of eluent system VII (60-100% B in 30 min). Following lyophilization the target peptide was isolated using a linear gradient of eluent system IV (50-70% B in 30 min).

UPLC (column 1, Eluent system I, 50-80% in 15 min) : R_t = 4.87 min.

ESI-MS m/z: 1474.8 [M+2H]$^{2+}$, 983.5 [M+3H+]$^{3+}$.

HR-MS (ESI): calcd. for $C_{140}H_{221}N_{29}O_{40}$, ([M+2H]$^{2+}$): 1474.3070, found: 1474.3063; calcd. for $C_{140}H_{222}N_{29}O_{40}$, ([M+3H]$^{3+}$): 983.2071, found: 983.2079.

E3((DEACM)$_2$^{butenyl}) b

$C_{142}H_{219}N_{29}O_{44}$
3036.5 g/mol
32

The peptide was synthesized at a 25 µmolar scale on a Rink amid resin. The first 9 aas were coupled by automated SPPS (**SOP1**). All following aas were added by manual SPPS (**SOP3** and **SOP4**) and after the final Fmoc deprotection the *N*-terminus was acylated (**SOP5**). After thorough washing of the resin Grubbs-catalyzed macrocyclization was performed (**SOP6**). The peptide was cleaved (**SOP7**) and after lyophilization the crude product was purified by semi-preparative HPLC utilizing column 3 with a linear gradient of eluent system IV (60-80% B in 30 min). Two peaks with masses corresponding to the target peptide were isolated.

UPLC (column 1, eluent system I, 40-80% in 15 min) : R_t = 8.5 min, 8.8 min.

ESI-MS *m/z*: 1518.8 $[M+2H]^{2+}$, 1012.9 $[M+3H+]^{3+}$.

HR-MS (ESI): calcd. for $C_{142}H_{221}N_{29}O_{44}$, ($[M+2H]^{2+}$): 1518.2968, found: 1518.2966; calcd. for $C_{142}H_{221}N_{29}O_{44}$, ($[M+3H]^{3+}$): 1012.5336, found: 1012.5337.

K3Sx

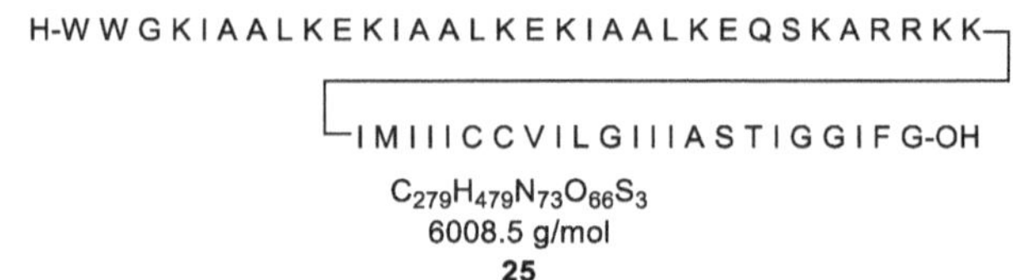

The peptide was synthesized at a 50 µmolar scale by automated SPPS (**SOP2**) following the protocol for peptides longer than 30 aas. After cleavage (**SOP7**) and lyophilization the crude product was purified by preparative HPLC utilizing column 5 with a linear gradient of eluent system VI (70-95% B in 30 min).

UPLC (column 1, Eluent system II, 70-99% in 8 min) : R_t = 6.25 min.

ESI-MS *m/z*: 1202.5 $[M+5H]^{5+}$, 1002.4 $[M+6H]^{6+}$.

HR-MS (ESI): calcd. for $C_{279}H_{484}N_{73}O_{66}S_3$, ($[M+5H]^{5+}$): 1201.9179 found: 1201.9201; calcd. for $C_{279}H_{485}N_{73}O_{66}S_3$, ($[M+6H]^{6+}$): 1001.7661, found: 1001.7655.

E3Syb

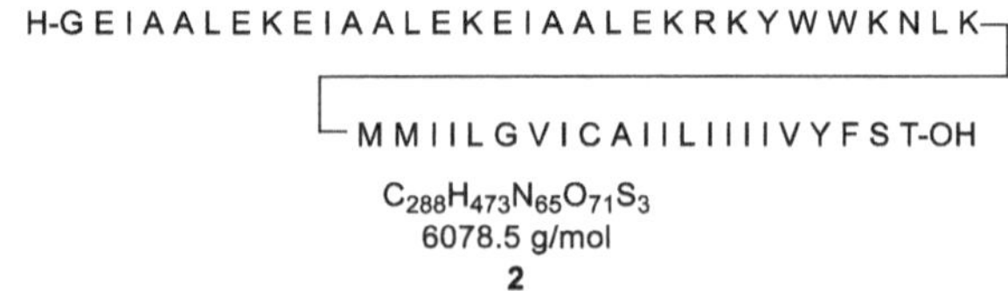

The peptide was synthesized at a 50 µmolar scale by automated SPPS (**SOP2**) following the protocol for peptides longer than 30 aas. After cleavage (**SOP7**) and lyophilization the crude product was purified by preparative HPLC utilizing column 5 with a linear gradient of eluent system VI (80-100% B in 30 min).

UPLC (column 1, Eluent system II, 80-99% in 8 min) : R_t = 6.40 min.

ESI-MS m/z: 1215.5 $[M+5H]^{5+}$, 1013.9 $[M+6H]^{6+}$.

HR-MS (ESI): calcd. for $C_{288}H_{478}N_{73}O_{65}S_3$, ($[M+5H]^{5+}$): 1215.8985 found: 1215.8985; calcd. for $C_{288}H_{479}N_{73}O_{65}S_3$, ($[M+6H]^{6+}$): 1013.4166, found: 1013.4154.

K3(NVOC)Sx

H-W W G K I A A L K E K I A A L K E K I A A L K E Q S K A R R K K—

—I M I I I C C V I L G I I I A S T I G G I F G-OH

$C_{289}H_{488}N_{74}O_{72}S_3$
6447.7 g/mol
26

The peptide was synthesized at a 50 µmolar scale by automated SPPS (**SOP2**) following the protocol for peptides longer than 30 aas. Fmoc-Lys(NVOC)-OH was introduced manually (**SOP3**). After cleavage (**SOP7**) and lyophilization the crude product was purified by semi-preparative HPLC utilizing column 4 with a linear gradient of eluent system VI (70-95% B in 30 min).

UPLC (column 1, Eluent system II, 70-99% in 8 min) : R_t = 6.10 min.

ESI-MS m/z: 1250.5 $[M+5H]^{5+}$, 1042.1 $[M+6H]^{6+}$.

HR-MS (ESI): calcd. for $C_{289}H_{493}N_{74}O_{72}S_3$, ($[M+5H]^{5+}$): 1249.7265 found: 1249.7229; calcd. for $C_{289}H_{494}N_{74}O_{72}S_3$, ($[M+6H]^{6+}$): 1041.6066, found: 1041.6045.

K3(NVOC)₂Sx

H-W W G K I A A L K E K I A A L K E K I A A L K E Q S K A R R K K—

—I M I I I C C V I L G I I I A S T I G G I F G-OH

$C_{299}H_{497}N_{75}O_{78}S_3$
6486.9 g/mol
27

The peptide was synthesized at a 50 µmolar scale by automated SPPS (**SOP2**) following the protocol for peptides longer than 30 aas. Fmoc-Lys(NVOC)-OH was introduced manually (**SOP3**). After cleavage (**SOP7**) and lyophilization the crude product was purified by semi-preparative HPLC utilizing column 4 with a linear gradient of eluent system VI (70-95% B in 30 min).

UPLC (column 1, Eluent system II, 70-99% in 8 min) : R_t = 6.80 min.

ESI-MS *m/z*: 1298.1 $[M+5H]^{5+}$, 1082.1 $[M+6H]^{6+}$.

HR-MS (ESI): calcd. for $C_{279}H_{484}N_{73}O_{66}S_3$, ($[M+5H]^{5+}$): 1297.5351 found: 1297.5342; calcd. for $C_{279}H_{485}N_{73}O_{66}S_3$, ($[M+6H]^{6+}$): 1081.4471, found: 1081.4301.

E3(DMNPB)₂Syb

H-G E I A A L E K E I A A L E K E I A A L E K R K Y W W K N L K—

—M M I I L G V I C A I I L I I I I V Y F S T-OH

$C_{312}H_{503}N_{67}O_{79}S_3$
6553.0 g/mol
28

The peptide was synthesized at a 50 µmolar scale by automated SPPS (**SOP2**) following the protocol for peptides longer than 30 aas. Fmoc-Glu(DMNPB)-OH was introduced manually (**SOP3**). After cleavage (**SOP7**) and lyophilization the crude product was purified by semi-preparative HPLC utilizing column 4 with a linear gradient of eluent system VI (80-100% B in 30 min).

UPLC (column 1, Eluent system II, 80-99% in 8 min) : $R_t = 5.90$ min.

ESI-MS *m/z*: 1311.5 $[M+5H]^{5+}$, 1093.1 $[M+6H]^{6+}$.

HR-MS (ESI): calcd. for $C_{312}H_{503}N_{67}O_{79}S_3$, ($[M+5H]^{5+}$): 1310.7386 found: 1310.7420; calcd. for $C_{288}H_{479}N_{73}O_{65}S_3$, ($[M+6H]^{6+}$): 1092.4500, found: 1092.4503.

K3(DEACM)₂Sx

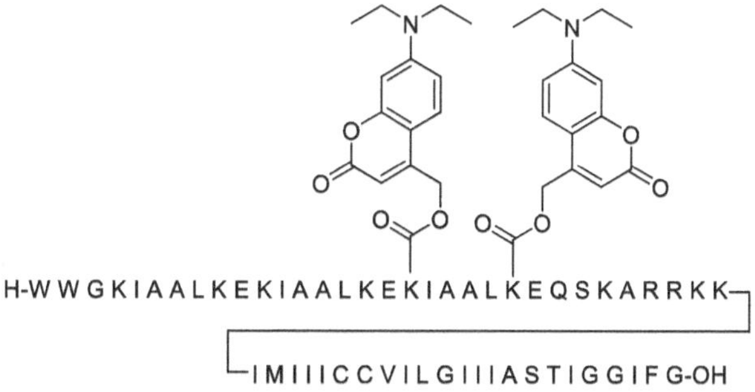

$C_{309}H_{509}N_{75}O_{74}S_3$
6555,1 g/mol
29

The peptide was synthesized at a 50 µmolar scale by automated SPPS (**SOP2**) following the protocol for peptides longer than 30 aas. Fmoc-Lys(DEACM)-OH was coupled manually (**SOP3**). After cleavage (**SOP7**) and lyophilization the crude product was purified by preparative HPLC utilizing column 5 with a linear gradient of eluent system VI (80-100% B in 30 min).

UPLC (column 1, Eluent system II, 70-99% in 8 min) : $R_t = 7.10$ min.

ESI-MS *m/z*: 1312.0 $[M+5H]^{5+}$, 1093.5 $[M+6H]^{6+}$.

HR-MS (ESI): calcd. for $C_{309}H_{514}N_{75}O_{74}S_3$, ($[M+5H]^{5+}$): 1310.1580 found: 1310.1589; calcd. for $C_{309}H_{515}N_{75}O_{74}S_3$, ($[M+6H]^{6+}$): 1092.7995, found: 1092.7979.

K3((DEACM)$_2$^{butenyl})Sx

H-W W G K I A A L K E K I A A L K E K I A A L K E Q S K A R R K K—

—I M I I I C C V I L G I I I A S T I G G I F G-OH

$C_{313}H_{513}N_{75}O_{74}S_3$
6607.2 g/mol
49

The peptide was synthesized at a 50 µmolar scale on a preloaded Wang resin. All aas except for Fmoc-Lys(DEACM^{allyl})-OH were coupled by automated SPPS (**SOP2**). The caged lysine was introduced by manual SPPS (**SOP3**). *N*-terminal Fmoc was kept on the peptide for the following metathesis. After thorough washing and drying of the resin Grubbs-catalyzed macrocyclization was performed on 100 mg of resin (**SOP6**). The peptide was cleaved (**SOP7**) and after lyophilization the crude product was purified by two-step HPLC. The bulk of impurities was removed on column 5 with a linear gradient of eluent V (70-100% B in 15 min) and the target peptide was isolated from column 3 with a linear gradient of eluent system V (75-90% B in 30 min). Fractions containing the product were identified by LC-MS.

LC-MS: R_t = 6.39 min.

ESI-MS *m/z*: 1322.4 [M+5H]$^{5+}$, 1102.1 [M+6H]$^{6+}$, 944.9 [M+7H]$^{7+}$.

HR-MS (ESI): calcd. for $C_{313}H_{513}N_{75}O_{74}S_3$, ([M+5H]$^{5+}$): 1321.56, found: 1321.55; calcd. for $C_{315}H_{513}N_{75}O_{74}S_3$, ([M+7H]$^{7+}$): 944.4055, found: 944.4047.

NY-ESO-1_long

H-L Q Q L S L L M W I T Q C F L-OH
$C_{86}H_{137}N_{19}O_{21}S_2$
1837.3 g/mol
51

The peptide was synthesized at a 50 µmolar scale on a preloaded Wang resin (**SOP1**). 50 mg of the on-resin peptide were reserved for the synthesis of NY-ESO-1_long_biotin. After cleavage (**SOP7**) additional washing steps with Et$_2$O (5 x 40 mL in total) were executed and the peptide was lyophilized. As the peptide is highly prone to aggregation in all tested solvents no HPLC purification or analysis could be performed.

ESI-MS *m/z*: 1838.0 [M+2H]$^{2+}$, 930.5 [M+Na+2H]$^{3+}$.

HR-MS (ESI): calcd. for $C_{86}H_{139}N_{18}NaO_{21}S_2$, ([M+2H]$^+$): 1836.9751, found: 1836.9702; calcd. for $C_{86}H_{140}N_{18}NaO_{21}S_2$, ([M+Na+2H]$^{3+}$): 929.9821, found: 929.9808.

NY-ESO1_short_pra5

H-S L L M {pra} I T Q V-OH

$C_{45}H_{78}N_{10}O_{13}S$

999.2

52

The peptide was built up at a 50 µmolar scale on a preloaded Wang resin (**SOP1**). After cleavage (**SOP7**) additional washing steps with Et$_2$O (5 x 40 mL in total) were executed and the peptide was lyophilized. As the peptide is highly prone to aggregation in all tested solvents no HPLC purification or analysis could be performed.

ESI-MS *m/z*: 999.5 [M+H]$^+$, 511.3 [M+Na+H]$^{2+}$.

HR-MS (ESI): calcd. for $C_{45}H_{79}N_{10}O_{13}S$, ([M+H]$^+$): 999.5543, found: 999.5540; calcd. for $C_{45}H_{79}N_{10}NaO_{13}S$, ([M+2H]$^{2+}$): 511.2718, found: 511.2717.

NY-ESO1_long_pra5

H-L Q Q L S L L M {pra}T Q C F L-OH

$C_{80}H_{132}N_{18}O_{21}S_2$

1746.2 g/mol

3

The peptide was built up at a 50 µmolar scale on a preloaded Wang resin (**SOP1**). After cleavage (**SOP7**) additional washing steps with Et$_2$O (5 x 40 mL in total) were executed and the peptide was lyophilized. As the peptide is highly prone to aggregation in all tested solvents no HPLC purification or analysis could be performed.

ESI-MS *m/z*: 1746.9 [M+H]$^+$, 873.5 [M+2H]$^{2+}$.

HR-MS (ESI): calcd. for $C_{80}H_{133}N_{18}O_{21}S_2$, ([M+H]$^+$): 1745.9329, found: 1745.9332; calcd. for $C_{80}H_{134}N_{18}O_{21}S_2$, ([M+2H]$^{2+}$): 873.4701, found: 873.4707.

NY-ESO1_long_pra8

H-L Q Q L S L LM W I T {pra} C F L-OH

$C_{86}H_{134}N_{18}O_{20}S_2$

1804.2 g/mol

56

The peptide was built up at a 50 µmolar scale on a preloaded Wang resin (**SOP1**). 50 mg of the on-resin peptide were reserved for the synthesis of NY-ESO-1_long_8_biotin. After cleavage (**SOP7**) additional washing steps with Et_2O (5 x 40 mL in total) were executed and the peptide was lyophilized. As the peptide is highly prone to aggregation in all tested solvents no HPLC purification or analysis could be performed.

ESI-MS *m/z*: 921.9 $[M+K+H]^{2+}$, 614.6 $[M+K+2H]^{3+}$.

HR-MS (ESI): calcd. for $C_{86}H_{135}N_{18}O_{20}S_2K_1$, ($[M+K+H]^{2+}$): 921.4584, found: 921.4554; calcd. for $C_{86}H_{136}N_{18}O_{20}S_2K_1$, ($[M+K+2H]^{3+}$): 614.6413, found: 614.6386.

Biotin_NY-ESO-1_long

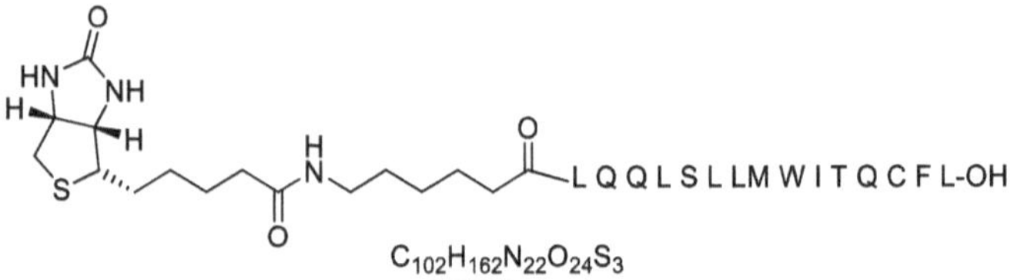

$C_{102}H_{162}N_{22}O_{24}S_3$
2176.7 g/mol
57

50 mg (~ 8 µmol) of the resin-bound peptide NY-ESO-1_long (**51**) were manually coupled with Fmoc-amino-caproic acid (**SOP4**). Following the removal of Fmoc, biotin was added analogous to the cystein coupling conditions (**SOP3**). After cleavage (**SOP7**), additional washing steps with Et_2O (5 x 20 mL in total) were executed and the peptide was lyophilized. As the peptide is highly prone to aggregation in all tested solvents no HPLC purification or analysis could be performed.

ESI-MS *m/z*: 1111.1 $[M+2Na]^{2+}$, 748.4 $[M+3Na]^{3+}$.

HR-MS (ESI): calcd. for $C_{102}H_{161}N_{21}O_{23}S_3$, ($[M+2H]^{+}$): 1099.5630, found: 1099.5641; calcd. for $C_{102}H_{162}N_{21}O_{23}S_3$, ($[M+Na+2H]^{3+}$): 733.3777, found: 733.3744.

Biotin_NY-ESO-1_long_pra8

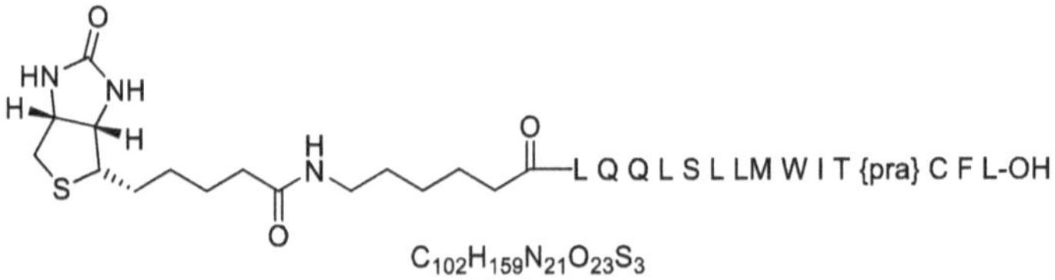

$C_{102}H_{159}N_{21}O_{23}S_3$
2143.7 g/mol
58

50 mg (~ 8 µmol) of the resin-bound peptide NY-ESO-1_long_pra8 (**56**) were manually coupled with Fmoc-amino-caproic acid (**SOP4**). Following the removal of Fmoc, biotin was added analogous to the cystein coupling conditions (**SOP3**). After cleavage (**SOP7**), additional washing steps with Et$_2$O (5 x 20 mL in total) were executed and the peptide was lyophilized. As the peptide is highly prone to aggregation in all tested solvents no HPLC purification or analysis could be performed.

ESI-MS *m/z*: 1072.7 [M+2H]$^{2+}$, 715.5 [M+3H]$^{3+}$.

HR-MS (ESI): calcd. for C$_{102}$H$_{161}$N$_{21}$O$_{23}$S$_3$, ([M+2H]$^+$): 1072.0613, found: 1072.0608; calcd. for C$_{102}$H$_{162}$N$_{21}$O$_{23}$S$_3$, ([M+3H]$^{3+}$): 715.0433, found: 715.0415.

Atto647-NY-ESO1_long_pra8

C$_{120}$H$_{176}$N$_{20}$O$_{24}$S$_3$
2379,0 g/mol
59

10 mg (~ 1.6 µmol) of the resin-bound peptide NY-ESO-1_long_8 was added to a 1.5 mL Eppendorf tube and flushed with argon. DIPEA (2.8 µL, 16 µmol, 10 eq) in dry and degassed DMF (100 µL) was added to the resin under argon stream and the resin was gently agitated at rt for 1 h. The Atto647-NHS-ester (1 mg, 1.23 µmol, 0.7 eq with respect to the peptide) in dry DMF (100 µL) was added to the resin and gently agitated in the dark for 12 h. The resin was transferred to a BD syringe, washed thoroughly (DMF, DCM, DMF 3x2 mL each, DCM 7x2 mL) and dried *in vacuo*. After cleavage (**SOP7**) and lyophilization the crude product was purified by HPLC utilizing column 3 with a linear gradient of eluent system V (75-100% B in 30 min, 55 °C). The product peak was identified by following the absorption at 647 nm. At the time of synthesis and purification, the molecular formula of Atto 647 was not yet published, thus, no HR-MS analysis was performed.

HPLC: (column 3, eluent system V, 75-100% B in 30 min, 1 mL/min) R_t = 17.7 min.

ESI-MS *m/z*: 1190.1 [M+2H]$^+$, 1201.1 [M+H+Na]$^{2+}$, 1209.1 [M+H+K]$^{2+}$.

gp100_long

H-V T H T Y L E P G P V T A Q V V L-OH
C$_{84}$H$_{134}$N$_{20}$O$_{25}$
1824.1 g/mol
60

The peptide was built up at a 50 μmolar scale on a preloaded Wang resin (**SOP1**). After cleavage (**SOP7**) and lyophilization the crude product was purified by preparative HPLC on column 5 with a linear gradient of eluent system IV (40-80% B in 30 min).

UPLC: (column 1, eluent system II, 20-90% B in 15 min, 0.40 mL/min) R_t = 10.94 min.

ESI-MS m/z: 913.0 [M+2H]$^+$, 609.0 [M+3H]$^{2+}$.

HR-MS (ESI): calcd. for $C_{84}H_{136}N_{20}O_{25}$, ([M+2H]$^{2+}$): 912.4987, found: 912.4990; calcd. for $C_{84}H_{137}N_{20}O_{25}$, ([M+3H]$^{3+}$): 608.6682, found: 608.6689.

gp100_short_5

H-Y L E P {pra} P V T A-OH

$C_{47}H_{69}N_9O_{14}$

984.1 g/mol

61

The peptide was built up at a 50 μmolar scale on a preloaded Wang resin (**SOP1**). After cleavage (**SOP7**) and lyophilization the crude product was purified by preparative HPLC on column 5 with a gradient of eluent system IV (30-70% B in 30 min).

UPLC: (column 1, eluent system II, 20-90% B in 15 min, 0.40 mL/min) R_t = 7.45 min.

ESI-MS m/z: 984.5 [M+H]$^+$, 503.7 [M+Na+H]$^{2+}$.

HR-MS (ESI): calcd. for $C_{47}H_{70}N_9O_{14}$, ([M+H]$^+$): 984.5037, found: 984.5034; calcd. for $C_{47}H_{71}N_9NaO_{14}$, ([M+Na+H]$^{2+}$): 503.7464, found: 503.7453.

gp100_long_pra5

H-V T H T Y L E P {pra} P V T A Q V V L-OH

$C_{87}H_{136}N_{20}O_{25}$

1862.2 g/mol

62

The peptide was built up at a 50 μmolar scale on a preloaded Wang resin (**SOP1**). After cleavage (**SOP7**) and lyophilization the crude product was purified by preparative HPLC on column 5 with a linear gradient of eluent system IV (50-90% B in 30 min).

UPLC: (column 1, eluent system II, 20-90% B in 15 min, 0.40 mL/min) R_t = 11.36 min.

ESI-MS m/z: 1863.2 [M+H]$^+$, 932.2 [M+2H]$^{2+}$.

HR-MS (ESI): calcd. for $C_{87}H_{137}N_{20}O_{25}$, ([M+H]$^+$): 1862.0058, found: 1862.0047; calcd. for $C_{87}H_{138}N_{20}O_{25}$, ([M+2H]$^{2+}$): 931.5066, found: 931.5065.

gp100_short_pra7

H-Y L E P G P {pra} T A-OH

$C_{44}H_{63}N_9O_{14}$
942.0 g/mol
63

The peptide was built up at a 50 µmolar scale on a preloaded Wang resin (**SOP1**). After cleavage (**SOP7**) and lyophilization the crude product was purified by preparative HPLC on column 5 with a gradient of eluent system IV (30-70% B in 30 min).

UPLC: (column 1, eluent system II, 20-90% B in 15 min, 0.40 mL/min) R_t = 5.63 min.

ESI-MS *m/z*: 942.5 $[M+H]^+$, 482.7 $[M+Na+H]^{2+}$.

HR-MS (ESI): calcd. for $C_{44}H_{64}N_9O_{14}$, ($[M+H]^+$): 942.4567, found: 942.4571; calcd. for $C_{44}H_{65}N_9O_{14}$, ($[M+2H]^{2+}$): 471.7320, found: 471.7324.

gp100_long_pra7

H-V T H T Y L E P G P {pra} T A Q V V L-OH

$C_{84}H_{130}N_{20}O_{25}$
1820.1 g/mol
64

The peptide was built up at a 50 µmolar scale on a preloaded Wang resin (**SOP1**). After cleavage (**SOP7**) and lyophilization the crude product was purified by preparative HPLC on column 5 with a linear gradient of eluent system IV (50-90% B in 30 min).

UPLC: (column 1, eluent system II, 20-90% B in 15 min, 0.40 mL/min) R_t = 10.44 min.

ESI-MS *m/z*: 1821.2 $[M+H]^+$, 910.7 $[M+2H]^{2+}$.

HR-MS (ESI): calcd. for $C_{84}H_{131}N_{20}O_{25}$, ($[M+H]^+$): 1819.9581, found: 1819.9581; calcd. for $C_{84}H_{132}N_{20}O_{25}$, ($[M+2H]^{2+}$): 910.4831, found: 910.4834.

gp100_short_pra8

H-Y L E P G P V {pra} A-OH

$C_{45}H_{65}N_9O_{13}$
940.1 g/mol
65

The peptide was built up at a 50 µmolar scale on a preloaded Wang resin (**SOP1**). After cleavage (**SOP7**) and lyophilization the crude product was purified by preparative HPLC on column 5 with a gradient of eluent system IV (30-70% B in 30 min).

UPLC: (column 1, eluent system II, 20-90% B in 15 min, 0.40 mL/min) R_t = 7.03 min.

ESI-MS *m/z*: 940.4 $[M+H]^+$, 489.7 $[M+K+H]^{2+}$.

HR-MS (ESI): calcd. for $C_{45}H_{66}N_9O_{13}$, ([M+H]$^+$): 940.4775, found: 940.4778; calcd. for $C_{45}H_{66}N_9NaO_{13}$, ([M+Na+H]$^{2+}$): 481.7333, found: 481.7324.

gp100_long_pra8

H-V T H T Y L E P G P V {pra} A Q V V L-OH

$C_{85}H_{132}N_{20}O_{24}$

1818.1

66

The peptide was built up at a 50 µmolar scale on a preloaded Wang resin (**SOP1**). After cleavage (**SOP7**) and lyophilization the crude product was purified by preparative HPLC on column 5 with a linear gradient of eluent system IV (50-90% B in 30 min).

UPLC: (column 1, eluent system II, 20-90% B in 15 min, 0.40 mL/min) R_t = 11.37 min.

ESI-MS *m/z*: 909.4 [M+2H]$^{2+}$, 614.3 [M+3H]$^{3+}$.

HR-MS (ESI): calcd. for $C_{85}H_{134}N_{20}O_{24}$, ([M+5H]$^{5+}$): 909.4934, found: 909.4947; calcd. for $C_{85}H_{135}N_{20}O_{24}$, ([M+2H]$^{2+}$): 606.6647, found: 606.6659.

Mart1_short_pra5

H-E A A G I {pra} I L T V-OH

$C_{45}H_{76}N_{10}O_{14}$

981.2 g/mol

67

The peptide was built up at a 50 µmolar scale on a preloaded Wang resin (**SOP1**). After cleavage (**SOP7**) and lyophilization the crude product was purified by preparative HPLC on column 5 with a gradient of eluent system IV (50-90% B in 30 min).

UPLC: (column 1, eluent system II, 20-90% B, 0.40 mL/min) R_t = 11.81 min.

ESI-MS *m/z*: 981.7 [M+H]$^+$, 502.4 [M+Na+H]$^{2+}$.

HR-MS (ESI): calcd. for $C_{45}H_{77}N_{10}O_{14}$, ([M+H]$^+$): 981.5615, found: 981.5608; calcd. for $C_{45}H_{77}N_{10}NaO_{14}$, ([M+Na+H]$^{2+}$): 502.2754, found: 502.2752.

Mart1_long_pra5

H-T T A E E A A G I {pra} I L T V I L G V-OH

$C_{80}H_{136}N_{18}O_{26}$

1766.1 g/mol

68

The peptide was built up at a 50 µmolar scale on a preloaded Wang resin (**SOP1**). After cleavage (**SOP7**) and lyophilization the crude product was purified by preparative HPLC on column 5 with a gradient of eluent system VII (30-100% B in 30 min).

Analytical HPLC: (column 2, eluent system VII, 40-100% B in 30 min, 1.0 mL/min) R_t = 20.89 min.

ESI-MS *m/z*: 1766.8 [M+H]$^+$, 883.4 [M+2H]$^{2+}$.

HR-MS (ESI): calcd. for $C_{80}H_{137}N_{18}O_{26}$, ([M+H]$^+$): 1765.9946, found: 1765.9932; calcd. for $C_{80}H_{138}N_{18}O_{26}$, ([M+2H]$^{2+}$): 883.5009, found: 883.5011.

Mart1_short_pra8

H-E A A G I G I {pra} T V-OH

$C_{41}H_{68}N_{10}O_{14}$
925.1 g/mol
69

The peptide was built up at a 50 µmolar scale on a preloaded Wang resin (**SOP1**). After cleavage (**SOP7**) and lyophilization the crude product was purified by preparative HPLC on column 5 with a gradient of eluent system IV (40-90% B in 30 min).

UPLC: (column 1, eluent system II, 20-90% B, 0.40 mL/min) R_t = 9.31 min.

ESI-MS *m/z*: 1165.8 [M+H]$^+$, 583.5 [M+2H]$^{2+}$.

HR-MS (ESI): calcd. for $C_{309}H_{515}N_{75}O_{74}S_3$, ([M+5H]$^{5+}$): 1165.6074, found: 1165.6071; calcd. for $C_{309}H_{516}N_{75}O_{74}S_3$, ([M+2H]$^{2+}$): 583.3074, found: 583.3075.

Mart1_long_pra8

H-T T A E E A A G I G I {pra} T V I L G V-OH

$C_{76}H_{128}N_{18}O_{26}$
1710,0
70

The peptide was built up at a 50 µmolar scale on a preloaded Wang resin (**SOP1**). After cleavage (**SOP7**) and lyophilization the crude product was purified by preparative HPLC on column 5 with a gradient of eluent system VII (30-100% B in 30 min).

Analytical HPLC: (column 2, eluent system VII, 40-100% B in 30 min, 1.0 mL/min) R_t = 15.95 min.

ESI-MS *m/z*: 1732.9 [M+Na]$^+$, 866.5 [M+Na+H]$^{2+}$.

HR-MS (ESI): calcd. for $C_{76}H_{128}N_{18}NaO_{26}$, ([M+Na]$^+$): 1731.9139 , found: 1731.9135; calcd. for $C_{76}H_{129}N_{18}NaO_{26}$, ([M+Na+H]$^{2+}$): 866.4606, found: 866.4613.

Mart1_short_pra9

H-E A A G I G I L {pra} V-OH

$C_{43}H_{72}N_{10}O_{13}$
937.1 g/mol
71

The peptide was built up at a 50 µmolar scale on a preloaded Wang resin (**SOP1**). After cleavage (**SOP7**) and lyophilization the crude product was purified by preparative HPLC on column 5 with a gradient of eluent system IV (50-90% B in 30 min).

UPLC: (column 1, eluent system II, 20-90% B, 0.40 mL/min) R_t = 11.54 min.

ESI-MS m/z: 937.5 $[M+H]^+$, 480.2 $[M+Na+H]^{2+}$.

HR-MS (ESI): calcd. for $C_{43}H_{73}N_{10}O_{13}$, ($[M+H]^+$): 937.5353, found: 937.5351; calcd. for $C_{43}H_{74}N_{10}O_{13}$, ($[M+2H]^{2+}$): 469.2713, found: 469.2712.

Mart1_long_22-39_9

H-T T A E E A A G I G I L {pra} V I L G V-OH

$C_{78}H_{132}N_{18}O_{25}$
1722,0 g/mol
72

The peptide was built up at a 50 µmolar scale on a preloaded Wang resin (**SOP1**). After cleavage (**SOP7**) and lyophilization the crude product was purified by preparative HPLC on column 5 with a gradient of eluent system VII (30-100% B in 30 min).

Analytical HPLC: (column 2, eluent system VII, 40-100% B in 30 min, 1.0 mL/min) R_t = 19.96 min.

ESI-MS m/z: 1723.0 $[M+H]^+$, 880.5 $[M+K+H]^{2+}$.

HR-MS (ESI): calcd. for $C_{78}H_{133}N_{18}O_{25}$, ($[M+H]^+$): 1721.9684 , found: 1721.9671; calcd. for $C_{78}H_{134}N_{18}O_{25}$, ($[M+2H]^{2+}$): 861.4878, found: 861.4876.

TM1

H-A A A W C L Q Q L S L L M W I T Q C F L P V F L A W A A A-OH

$C_{157}H_{233}N_{35}O_{35}S_3$
3267.0 g/mol
73

The peptide was built up at a 50 µmolar scale on a preloaded Wang resin (**SOP1**). After cleavage (**SOP7**) and lyophilization the crude product was purified by preparative HPLC on column 5 with a gradient of eluent system VI (60-100% B in 30 min).

Analytical HPLC: (column 2, Eluent system VI, 50-100% B in 30 min, 1.0 mL/min) R_t = 15.27 min.

ESI-MS m/z: 1634.3 [M+2H]$^{2+}$, 1097.2 [M+Na+2H]$^{3+}$.

HR-MS (ESI): calcd. for $C_{157}H_{235}N_{35}O_{35}S_3$, ([M+2H]$^{2+}$): 1633.3418, found: 1633.3409; calcd. for $C_{157}H_{235}N_{35}NaO_{35}S_3$, ([M+Na+2H]$^{3+}$): 1096.5576, found: 1095.5575.

TM1-pra5

H-A A A W C L Q Q L S L L M {pra} I T Q C F L P V F L A W A A A-OH

$C_{151}H_{228}N_{34}O_{35}S_3$
3175.9 g/mol
74

The peptide was built up at a 50 µmolar scale on a preloaded Wang resin (**SOP1**). After cleavage (**SOP7**) and lyophilization the crude product was purified by preparative HPLC on column 5 with a gradient of eluent system VI (60-100% B in 30 min).

Analytical HPLC: (column 2, Eluent system VI, 50-100% B in 30 min, 1.0 mL/min) R_t = 18.20 min.

ESI-MS m/z: 1588.9 [M+2H]$^{2+}$, 1059.6 [M+3H]$^{3+}$.

HR-MS (ESI): calcd. for $C_{151}H_{230}N_{34}O_{35}S_3$, ([M+2H]$^{+}$): 1587.8207, found: 1587.8197; calcd. for $C_{151}H_{231}N_{34}O_{35}S_3$, ([M+3H]$^{3+}$): 1058.8829, found: 1058.8817.

TM1_azi7

H-A A A W C L Q Q L S L L M W I {azi} Q C F L P V F L A W A A A-OH

$C_{157}H_{232}N_{38}O_{34}S_3$
3292.0 g/mol
75

The peptide was built up at a 50 µmolar scale on a preloaded Wang resin (**SOP1**). After cleavage (**SOP7**) and lyophilization the crude product was purified by preparative HPLC on column 5 with a gradient of eluent system VI (70-100% B in 30 min, 50 °C).

Analytical HPLC: (column 2, Eluent system VI, 70-100% B in 60 min, 1.0 mL/min, 50 °C) R_t = 25.90 min.

ESI-MS m/z: 1657.8 [M+Na+H]$^{2+}$, 1100.8 [M+K+2H]$^{3+}$.

HR-MS (ESI): calcd. for $C_{157}H_{233}N_{35}O_{34}S_3Na$, ([M+Na+H]$^{2+}$): 1656.8360, found: 1656.8356; calcd. for $C_{158}H_{234}N_{35}O_{33}S_3Na$, ([M+Na+2H]$^{3+}$): 1104.8931, found: 1104.8913.

TM1_pra7

H-A A A W C L Q Q L S L L M W I {pra} Q C F L P V F L A W A A A-OH

$C_{158}H_{231}N_{35}O_{34}S_3$
3261.0 g/mol
76

The peptide was built up at a 50 µmolar scale on a preloaded Wang resin (**SOP1**). After cleavage (**SOP7**) and lyophilization the crude product was purified by preparative HPLC on column 5 with a gradient of eluent system VI (70-100% B in 30 min, 50 °C).

Analytical HPLC: (column 2, eluent system VI, 70-100% B in 60 min, 1.0 mL/min, 50 °C) R_t = 27.1 min.

ESI-MS m/z: 1630.8 [M+2H]$^{2+}$, 1100.2 [M+K+2H]$^{3+}$.

HR-MS (ESI): calcd. for $C_{158}H_{233}N_{35}O_{34}S_3$, ([M+2H]$^{2+}$): 1630.3365, found: 1630.3373; calcd. for $C_{158}H_{232}N_{35}O_{33}S_3Na$, ([M+Na+H]$^{2+}$): 1641.3275, found: 1641.3282; calcd. for $C_{158}H_{233}N_{35}O_{33}S_3K$, ([M+K+H]$^{3+}$): 1099.8787, found: 1099.8776.

TM1-pra8

H-A A A W C L Q Q L S L L M W I T {pra} C F L P V F L A W A A A-OH

$C_{157}H_{230}N_{34}O_{34}S_3$
3234.0 g/mol
77

The peptide was built up at a 50 µmolar scale on a preloaded Wang resin (**SOP1**). After cleavage (**SOP7**) and lyophilization the crude product was purified by preparative HPLC on column 5 with a gradient of eluent system VI (60-100% B in 30 min).

Analytical HPLC: (column 2, Eluent system VI, 50-100% B in 30 min, 1.0 mL/min) R_t = 22.6 min.

ESI-MS m/z: 1628.8 [M+Na+H]$^{2+}$, 1091.6 [M+K+2H]$^{3+}$.

HR-MS (ESI): calcd. for $C_{157}H_{232}N_{34}O_{34}S_3$, ([M+H]$^+$): 1616.8311, found: 1616.8308; calcd. for $C_{157}H_{233}N_{34}O_{34}S_3$, ([M+3H]$^{3+}$): 1078.2231, found: 1078.2219.

TM7_pra7

H-A A A W L L C L V V L S L L M W I {pra} Q C F L P V F W A A A-OH

$C_{161}H_{239}N_{33}O_{32}S_3$
3242.7 g/mol
78

The peptide was built up at a 50 µmolar scale on a preloaded Wang resin (**SOP1**). After cleavage (**SOP7**) and lyophilization the crude product was purified by preparative HPLC on column 5 with a gradient of eluent system VI (60-100% B in 30 min, 50 °C).

Analytical HPLC: (column 2, Eluent system VI, 50-100% B in 30 min, 1.0 mL/min) R_t = 15.3 min.

ESI-MS m/z: 1634.3 [M+2H]$^{2+}$, 1097.2 [M+Na+2H]$^{3+}$.

HR-MS (ESI): calcd. for $C_{161}H_{239}N_{33}O_{32}S_3$, ([M+2H]$^{2+}$): 1622.3698, found: 1622.3683; calcd. for $C_{161}H_{239}N_{33}O_{32}S_3$, ([M+Na+2H]$^{3+}$): 1096.5576, found: 1095.5575.

TM8_pra7

H-A A A W F V L L C L V V L S L L L M W I {pra} Q C F L P W A A A-OH

$C_{161}H_{239}N_{33}O_{32}S_3$
3245.1 g/mol
79

The peptide was built up at a 50 µmolar scale on a preloaded Wang resin (**SOP1**). After cleavage (**SOP7**) and lyophilization the crude product was purified by preparative HPLC on column 5 with a gradient of eluent system VI (60-100% B in 30 min).

Analytical HPLC: (column 2, Eluent system VI, 70-100% B in 60 min, 1.0 mL/min) R_t = 27.8 min.

ESI-MS m/z: 1634.3 [M+2H]$^{2+}$, 1097.2 [M+Na+2H]$^{3+}$.

HR-MS (ESI): calcd. for $C_{161}H_{239}N_{33}O_{32}S_3$, ([M+2H]$^{2+}$): 1633.3418, found: 1633.3409; calcd. for $C_{161}H_{239}N_{33}O_{32}S_3$, ([M+Na+2H]$^{3+}$): 1096.5576, found: 1095.5575.

TM9

H-A A A W P F V L L C L Q Q L S L L M W I T Q C F L W A A A-OH

$C_{160}H_{239}N_{35}O_{35}S_3$
3309.1 g/mol
80

The peptide was built up at a 50 µmolar scale on a preloaded Wang resin (**SOP1**). After cleavage (**SOP7**) and lyophilization the crude product was purified by preparative HPLC on column 5 with a gradient of eluent system VI (60-100% B in 30 min).

Analytical HPLC: (column 2, Eluent system VI, 50-100% B in 30 min, 1.0 mL/min) R_t = 16.5 min.

ESI-MS m/z: 1666.3 [M+Na+H]$^{2+}$, 1111.2 [M+Na+2H]$^{3+}$.

HR-MS (ESI): calcd. for $C_{160}H_{241}N_{35}O_{35}S_3$, ([M+2H]$^{2+}$): 1654.3653, found: 1654.3637; calcd. for $C_{160}H_{241}N_{35}NaO_{35}S_3$, ([M+Na+2H]$^{3+}$): 1110.5733, found: 1110.5705.

TM9-pra5

H-A A A W P F V L L C L Q Q L S L L M {pra} I T Q C F L W A A A-OH

$C_{154}H_{234}N_{34}O_{35}S_3$
3217.9 g/mol
81

The peptide was built up at a 50 μmolar scale on a preloaded Wang resin (**SOP1**). After cleavage (**SOP7**) and lyophilization the crude product was purified by preparative HPLC on column 5 with a gradient of eluent system VI (60-100% B in 30 min).

Analytical HPLC: (column 2, Eluent system VI, 50-100% B in 30 min, 1.0 mL/min) R_t = 19.7 min.

ESI-MS *m/z*: 1620.7 $[M+Na+H]^{2+}$, 1080.8 $[M+Na+2H]^{3+}$.

HR-MS (ESI): calcd. for $C_{154}H_{236}N_{34}O_{35}S_3$, $([M+2H]^+)$: 1608.8442, found: 1608.8426; calcd. for $C_{154}H_{236}N_{34}NaO_{35}S_3$, $([M+Na+2H]^{3+})$: 1080.2259, found: 1080.2279.

TM9-pra7

H-A A A W F V L L C L V V L S L L M W I {pra} Q C F L P W A A A-OH

$C_{161}H_{239}N_{33}O_{32}S_3$
3245.1 g/mol
82

The peptide was built up at a 50 μmolar scale on a preloaded Wang resin (**SOP1**). After cleavage (**SOP7**) and lyophilization the crude product was purified by preparative HPLC on column 5 with a gradient of eluent system VI (60-100% B in 30 min).

Analytical HPLC: (column 2, Eluent system VI, 70-100% B in 60 min, 1.0 mL/min) R_t = 27.2 min.

ESI-MS *m/z*: 1634.4 $[M+Na+H]^{2+}$, 1095.2 $[M+K+2H]^{3+}$.

HR-MS (ESI): calcd. for $C_{161}H_{240}N_{33}NaO_{32}S_3$, $([M+Na+H]^{2+})$: 1633.3614, found: 1633.3623; calcd. for $C_{161}H_{240}KN_{34}O_{34}S_3$, $([M+K+H]^{2+})$: 1641.3483, found: 1641.3490.

TM9-azi7

H-A A A W P F V L L C L V V L S L L M W I {azi} Q C F L W A A A-OH

$C_{160}H_{240}N_{36}O_{32}S_3$
3276.0 g/mol
83

The peptide was built up at a 50 μmolar scale on a preloaded Wang resin (**SOP1**). After cleavage (**SOP7**) and lyophilization the crude product was purified by preparative HPLC on column 5 with a gradient of eluent system VI (70-100% B in 30 min).

Analytical HPLC: (column 2, Eluent system VI, 50-100% B in 30 min, 1.0 mL/min) R_t = 21.2 min.

ESI-MS m/z: 1105.6 [M+K+2H]$^{3+}$.

HR-MS (ESI): calcd. for $C_{161}H_{242}N_{36}O_{32}S_3$, ([M+2H]$^{2+}$): 1637.8784, found: 1637.8789; calcd. for $C_{161}H_{243}N_{34}O_{34}S_3$, ([M+3H]$^{3+}$): 1092.2547, found: 1092.2565.

TM9-pra8

H-A A A W P F V L L C L Q Q L S L L M W I T {pra} C F L W A A A-OH

$C_{160}H_{236}N_{34}O_{34}S_3$
3276.0 g/mol
1

The peptide was built up at a 50 µmolar scale on a preloaded Wang resin (**SOP1**). After cleavage (**SOP7**) and lyophilization the crude product was purified by preparative HPLC on column 5 with a gradient of eluent system VI (60-100% B in 30 min).

Analytical HPLC: (column 2, Eluent system VI, 50-100% B in 30 min, 1.0 mL/min) R_t = 22.59 min.

ESI-MS m/z: 1649.7 [M+Na+H]$^{2+}$, 1100.1 [M+Na+2H]$^{3+}$.

HR-MS (ESI): calcd. for $C_{160}H_{238}N_{34}O_{34}S_3$, ([M+2H]$^{+}$): 1637.8545, found: 1637.8518; calcd. for $C_{160}H_{239}N_{34}O_{34}S_3$, ([M+3H]$^{3+}$): 1092.2388, found: 1092.2382.

ControlWW

H-A A A W L I I I I V I V V V L L G I {pra} L A L I I G L W A A A-OH

$C_{156}H_{258}N_{32}O_{31}$
3078,0 g/mol
84

The peptide was built up at a 50 µmolar scale on a preloaded Wang resin (**SOP1**). After cleavage (**SOP7**) and lyophilization the crude product was purified by preparative HPLC on column 5 with a gradient of eluent system VIII (75-100% B in 30 min, 50 °C).

Analytical HPLC: (column 2, Eluent system VIII, 75-100% B in 30 min, 1.0 mL/min, 50 °C) R_t = 15.27 min.

ESI-MS m/z: 1634.3 [M+2H]$^{2+}$, 1097.2 [M+3H]$^{3+}$.

HR-MS (ESI): calcd. for $C_{156}H_{260}N_{32}O_{31}$, ([M+2H]$^{2+}$): 1538.9871, found: 1538.9851; calcd. for $C_{156}H_{261}N_{32}O_{31}$, ([M+3H]$^{3+}$): 1026.3271, found: 1026.3265.

ICP47

H-M S W A L E M A D T F L D T M R V G P R T Y A D V R D E I N K R G R E-OH

$C_{176}H_{281}N_{53}O_{56}S_3$
4131.7
85

The peptide was built up at a 50 µmolar scale on a preloaded Wang resin (**SOP1**). After cleavage (**SOP7**) and lyophilization the crude product was purified by preparative HPLC on column 5 with a gradient of eluent system IV (50-90% B in 30 min).

UPLC: (column 1, eluent system II, 20-90% B in 15 min, 0.40 mL/min) R_t = 13.98 min.

ESI-MS *m/z*: 1033.8 $[M+4H]^{4+}$, 827.2 $[M+5H]^{5+}$.

HR-MS (ESI): calcd. for $C_{176}H_{285}N_{53}O_{56}S_3$, ($[M+4H]^{4+}$): 1033.2556, found: 1033.2568; calcd. for $C_{176}H_{286}N_{53}O_{56}S_3$, ($[M+5H]^{5+}$): 826.8059, found: 826.8069.

HA_long_clickable

H-Y G A C P K Y V {pra} Q N T L K L A T G M R N-OH
$C_{102}H_{163}N_{29}O_{29}S_2$
2323.7 g/mol
86

The peptide built up at a 50 µmolar scale on a preloaded Wang resin (**SOP1**). After cleavage (**SOP7**) and lyophilization the crude product was purified by preparative HPLC on column 5 with a gradient of eluent system IV 30-70% B in 30 min).

UPLC: (column 1, eluent system II, 20-90% B, 0.40 mL/min) R_t = 8.99 min.

ESI-MS *m/z*: 1162.6 $[M+2H]^{2+}$, 775.4 $[M+3H]^{3+}$.

HR-MS (ESI): calcd. for $C_{102}H_{166}N_{29}O_{29}S_2$, ($[M+3H]^{3+}$): 775.0610, found: 775.0613.

HA_long_non-clickable

H-Y G A C P K Y V K Q N T L K L A T G M R N-OH
$C_{103}H_{170}N_{30}O_{29}S_2$
2356.8 g/mol
89

The peptide built up at a 50 µmolar scale on a preloaded Wang resin (**SOP1**). After cleavage (**SOP7**) and lyophilization the crude product was purified by preparative HPLC on column 5 with a gradient of eluent system IV (30-70% B in 30 min).

UPLC: (Eluent system II, 20-90% B, 0.40 mL/min) R_t = 7.57 min.

ESI-MS *m/z*: 1079.0 $[M+2H]^{2+}$, 786.4 $[M+3H]^{3+}$.

HR-MS (ESI): calcd. for $C_{103}H_{172}N_{30}O_{29}S_2$, ([M+2H]$^{2+}$): 1178.6168, found: 1178.6183, calcd. for $C_{103}H_{173}N_{30}O_{29}S_2$, ([M+3H]$^{3+}$): 786.0803, found: 786.0813.

HA_short_non-clickable

H-P K Y V K Q N T L K KL A T-OH

$C_{69}H_{118}N_{18}O_{19}$
1503.8 g/mol
87

The peptide built up at a 50 µmolar scale on a preloaded Wang resin (**SOP1**). After cleavage (**SOP7**) and lyophilization the crude product was purified by preparative HPLC on column 5 with a gradient of eluent system IV (30-70% B in 30 min).

UPLC: (Eluent system II, 20-90% B, 0.40 mL/min) R_t = 6.65 min.

ESI-MS *m/z*: 752.4 [M+2H]$^{2+}$, 501.9 [M+3H]$^{3+}$.

HR-MS (ESI): calcd. for $C_{69}H_{120}N_{18}O_{19}$, ([M+2H]$^{2+}$): 752.4483, found: 752.4491, calcd. for $C_{69}H_{121}N_{18}O_{19}$, ([M+3H]$^{3+}$): 501.9680, found: 501.9687.

HA_short_clickable

H-P K Y V {pra} Q N T L K KL A T-OH

$C_{68}H_{111}N_{17}O_{19}$
1470.7 g/mol
88

The peptide built up at a 50 µmolar scale on a preloaded Wang resin (**SOP1**). After cleavage (**SOP7**) and lyophilization the crude product was purified by preparative HPLC on column 5 with a gradient of eluent system IV (30-70% B in 30 min).

UPLC: (Eluent system II, 20-90% B, 0.40 mL/min) R_t = 7.94 min.

ESI-MS *m/z*: 735.9 [M+2H]$^{2+}$, 490.9 [M+3H]$^{3+}$.

HR-MS (ESI): calcd. for $C_{68}H_{113}N_{17}O_{19}$, ([M+2H]$^{2+}$): 775.9191, found: 775.9198.

8 Appendix

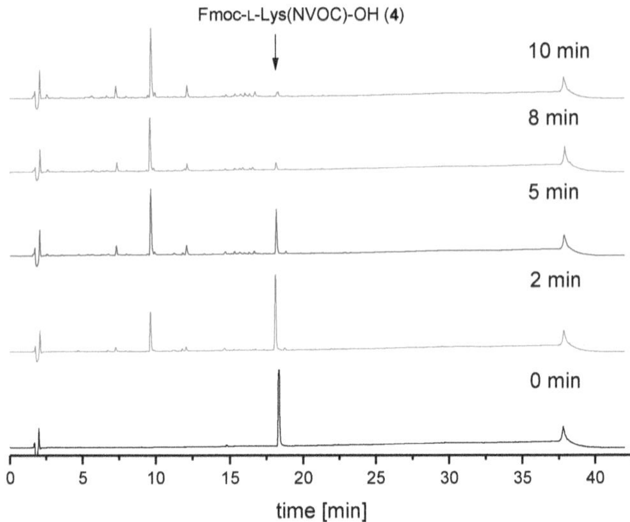

Figure-A 1: Uncaging of Fmoc-L-Lys(NVOC)-OH (4) (3.4 mM in MeOH) by method a (see section 7.1.5). Samples were drawn at the indicated timepoints and analyzed by analytical HPLC (20-90% B in 30 min, solvent system IV, flow 1 mL/min). Peaks were detected by absorption at 347 nm.

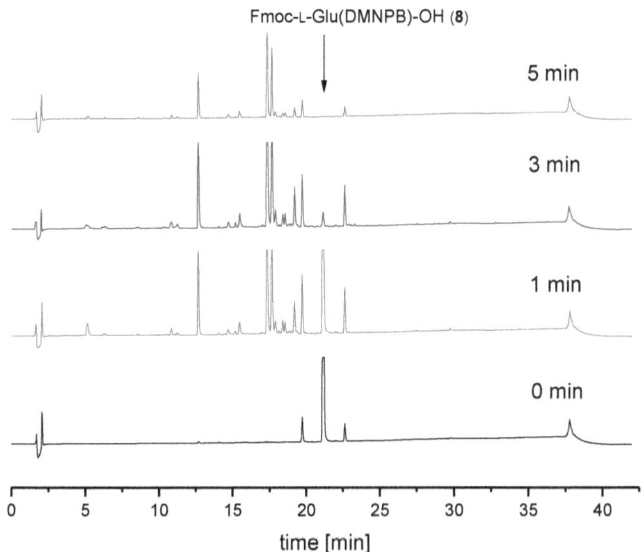

Figure-A 2: Uncaging of Fmoc-L-Glu(DMNPB)-OH (8) (3.4 mM in MeOH) by method a (see section 7.1.5). Samples were drawn at the indicated timepoints and analyzed by analytical HPLC (20-90% B in 30 min, solvent system IV, flow 1 mL/min). Peaks were detected by absorption at 347 nm.

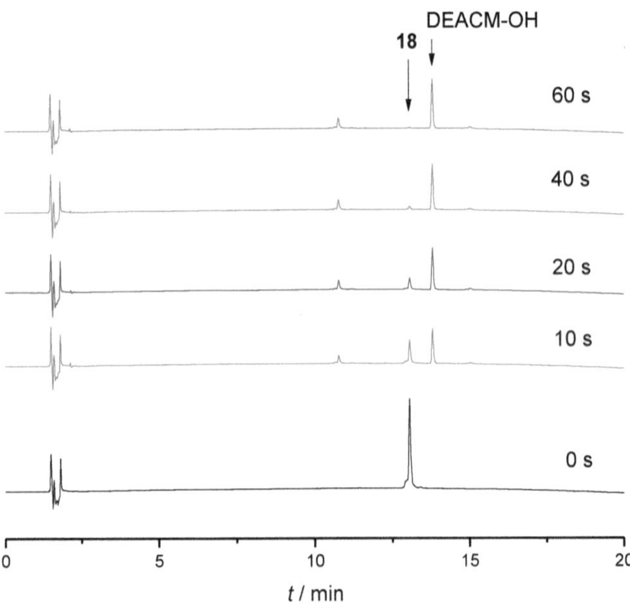

Figure-A 3: Uncaging of 18 (3.6 µM in HEPES buffer20 mM, pH 7.4) by method a (see section 7.1.5). Samples were drawn at the indicated timepoints and analyzed by UPLC (20-90% B in 30 min, solvent system IV, flow 1 mL/min). Peaks were detected by absorption at 390 nm

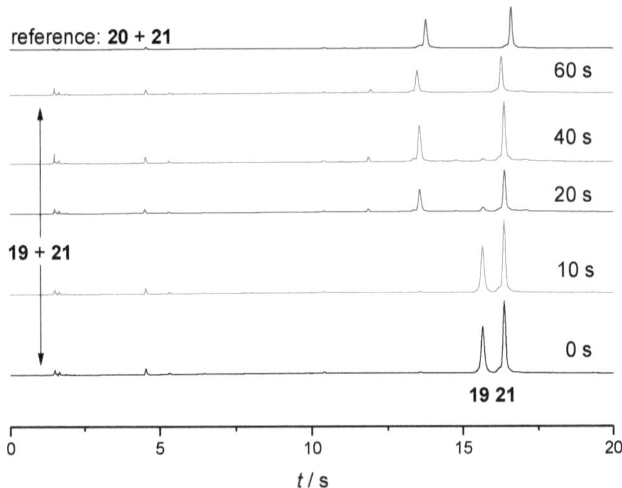

Figure-A 4: Uncaging of 19 in coiled coil complex with 21 (40 µM each in phosphate buffer 10 mM, pH 7) by method c (see section 7.1.5). Samples were drawn at the indicated timepoints and analyzed by UPLC (20-90% B in 20 min, solvent system II, flow 0.8 mL/min). Peaks were detected by absorption at 280 nm. As a reference, 20 and 21, which corresponds to the uncaged peptide combination was analyzed by UPLC with the same conditions.

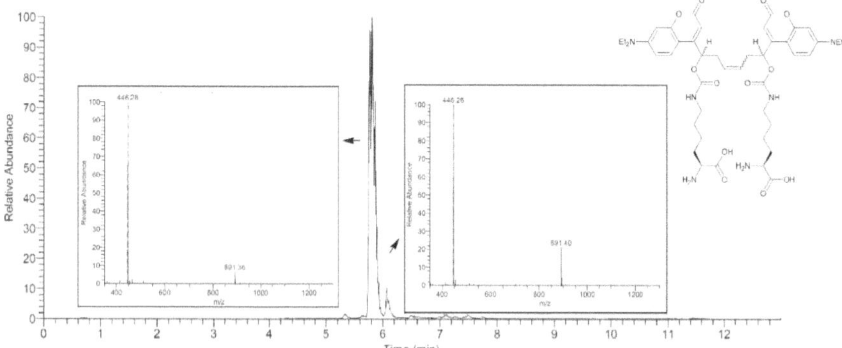

*Figure-A 5: LC-MS analysis of compound **40** (100 μM in ultrapure water) used to study uncaging behavior of the novel photocleavable staple at 0 min irradiation. The 50 μL sample was diluted with 10 μL of MeCN and 5 μL were injected for analysis. The chromatogram was recorded as total ion count normalized to the relative abundance. The separation was conducted with a linear gradient of 10 to 80% B in 8 min and subsequent isocratic elution with 100% B (see 7.3.4 for eluents). ESI-MS spectra were averaged over the width of the peaks. One main peak at 5.9 min and one less intense peak at 6.1 min with the same mass could be identified as stereo isomers of the educt.*

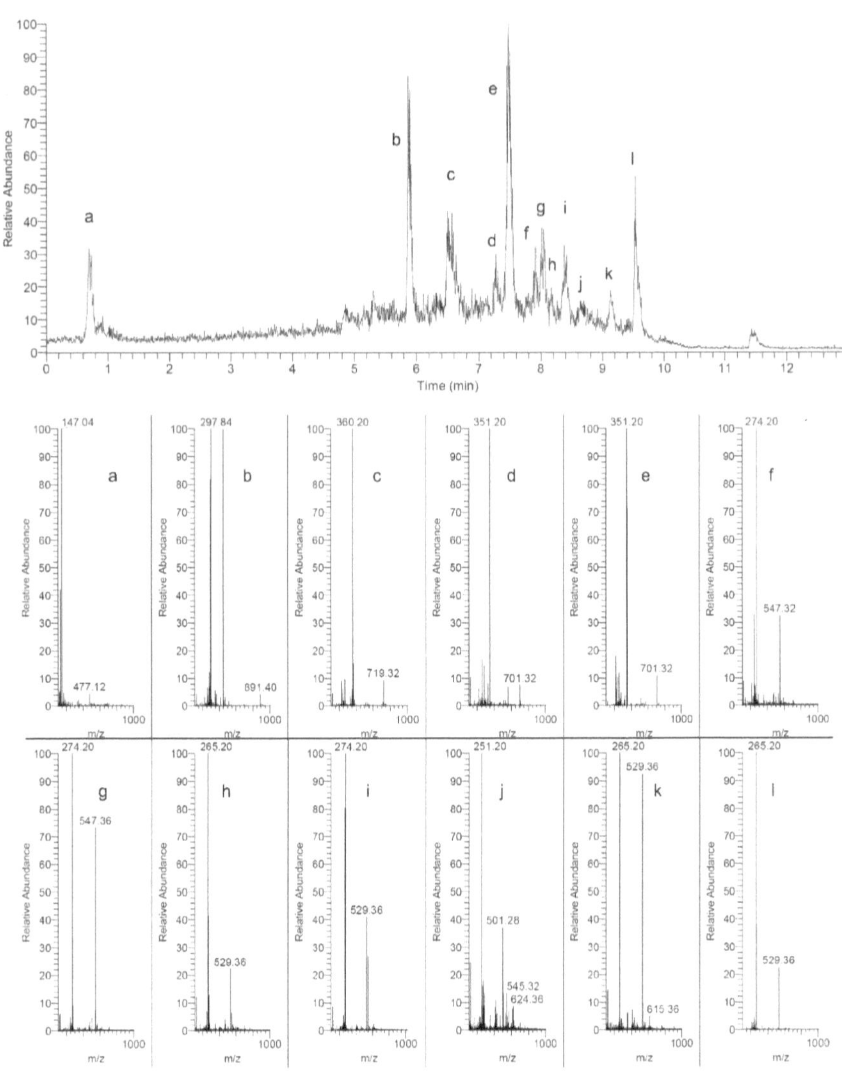

Figure-A 6: LC-MS analysis of compound **40** (100 μM in ultrapure water) used to study uncaging behavior of the novel photocleavable staple at 3 min irradiation by method b (see section 7.1.5). The 50 μL sample was diluted with 10 μL of MeCN and 5 μL were injected for analysis. The chromatogram was recorded as total ion count normalized to the relative abundance. The separation was conducted with a linear gradient of 10 to 80% B in 8 min and subsequent isocratic elution with 100% B (see 7.3.4 for eluents). ESI-MS spectra were averaged over the width of the peaks. One main peak at 5.9 min and one less intense peak at 6.1 min with the same mass could be identified as stereo isomers of the educt.

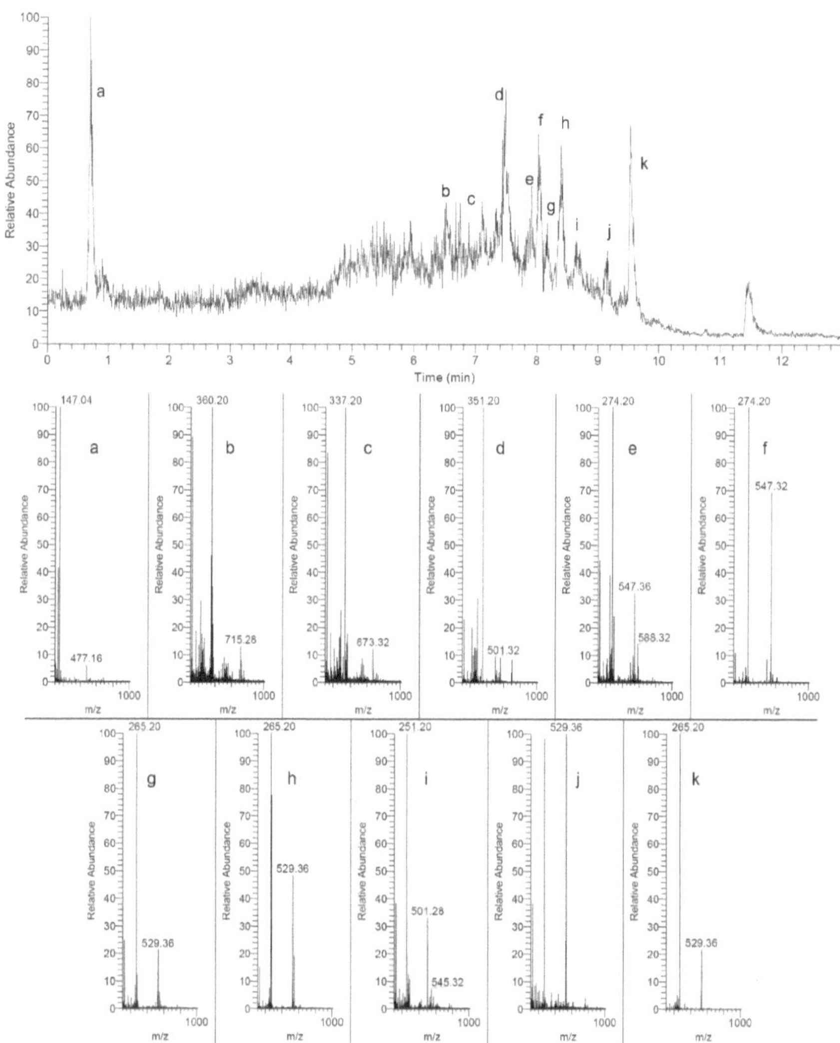

*Figure-A 7: LC-MS analysis of compound **40** (100 μM in ultrapure water) used to study uncaging behavior of the novel photocleavable staple at 30 min irratiation. The 50 μL sample was diluted with 10 μL of MeCN and 5 μL were injected for analysis. The chromatogram was recorded as total ion count normalized to the relative abundance. The separation was conducted with a linear gradient of 10 to 80% B in 8 min and subsequent isocratic elution with 100% B (see 7.3.4 for eluents). ESI-MS spectra were averaged over the width of the peaks. One main peak at 5.9 min and one less intense peak at 6.1 min with the same mass could be identified as stereo isomers of the educt.*

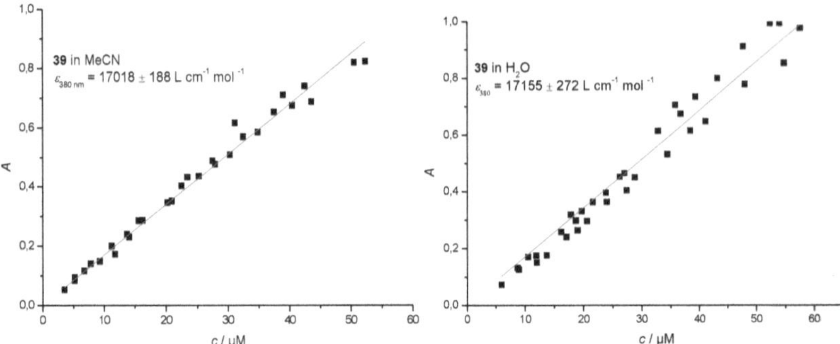

Figure-A 8: Measurement of the extinction coefficient of 39 at 380 nm in MeCN (left) and H₂O (right). A dilution series was prepared from three separately lyophilized and weighed aliquots and for each concentration three separate samples were prepared. The extinction coefficient was obtained as the slope of the linear fit of the data points.

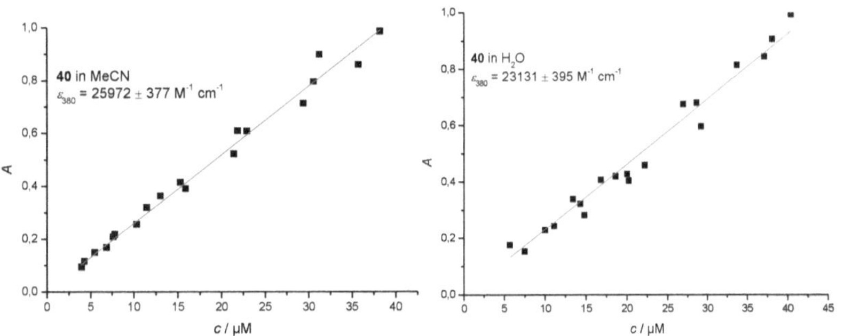

Figure-A 9: Measurement of the extinction coefficient of 40 at 380 nm in MeCN (left) and H₂O (right). A dilution series was prepared from three separately lyophilized and weighed aliquots and for each concentration three separate samples were prepared. The extinction coefficient was obtained as the slope of the linear fit of the data points.